# 797,885 Books

are available to read at

www.ForgottenBooks.com

Forgotten Books' App
Available for mobile, tablet & eReader

ISBN 978-1-331-95088-2
PIBN 10258671

This book is a reproduction of an important historical work. Forgotten Books uses state-of-the-art technology to digitally reconstruct the work, preserving the original format whilst repairing imperfections present in the aged copy. In rare cases, an imperfection in the original, such as a blemish or missing page, may be replicated in our edition. We do, however, repair the vast majority of imperfections successfully; any imperfections that remain are intentionally left to preserve the state of such historical works.

Forgotten Books is a registered trademark of FB &c Ltd.
Copyright © 2017 FB &c Ltd.
FB &c Ltd, Dalton House, 60 Windsor Avenue, London, SW19 2RR.
Company number 08720141. Registered in England and Wales.

For support please visit www.forgottenbooks.com

# 1 MONTH OF FREE READING

## at
## www.ForgottenBooks.com

By purchasing this book you are eligible for one month membership to ForgottenBooks.com, giving you unlimited access to our entire collection of over 700,000 titles via our web site and mobile apps.

To claim your free month visit:

www.forgottenbooks.com/free258671

\* Offer is valid for 45 days from date of purchase. Terms and conditions apply.

English
Français
Deutsche
Italiano
Español
Português

# www.forgottenbooks.com

**Mythology** Photography **Fiction**
Fishing Christianity **Art** Cooking
Essays Buddhism Freemasonry
Medicine **Biology** Music **Ancient Egypt** Evolution Carpentry Physics
Dance Geology **Mathematics** Fitness
Shakespeare **Folklore** Yoga Marketing
**Confidence** Immortality Biographies
Poetry **Psychology** Witchcraft
Electronics Chemistry History **Law**
Accounting **Philosophy** Anthropology
Alchemy Drama Quantum Mechanics
Atheism Sexual Health **Ancient History**
**Entrepreneurship** Languages Sport
Paleontology Needlework Islam
**Metaphysics** Investment Archaeology
Parenting Statistics Criminology
**Motivational**

# A TEXT-BOOK

OF

# MATERIA MEDICA AND PHARMACY

## FOR MEDICAL STUDENTS

BY

V. E. HENDERSON, M.A., M.B.

Associate Professor of Pharmacy and Pharmacology

IN THE UNIVERSITY OF TORONTO

UNIVERSITY PRESS
TORONTO

# PREFACE TO THE SECOND EDITION.

It was with the greatest reluctance that the authors of the first edition undertook in 1908 the preparation of another medical text-book. They found however that none of the books on Pharmacy and Materia Medica were at all suitable for use in their classes in the University of Toronto. Most of the books on these subjects contain also sections dealing with Pharmacology and Therapeutics. These sections are rarely accurate. Changes in the official Pharmacopœia occur much more slowly than does our knowledge of Pharmacology. Several of the better books on Pharmacy and Materia Medica are intended for students in the United States and deal largely with the U.S. Pharmacopœia. As there are several good books on Pharmacology on the market which undergo frequent revision, the authors did not include any material relating to this subject.

The classification of drugs according to their botanical, mineral or animal origin is no longer of importance, nor does a pharmacological classification furnish an arrangement useful for reference purposes. A use of one of these types of classification greatly mars some otherwise useful books.

These considerations forced the authors to compile this text-book and they hope that they have succeeded in placing before the student a book which will aid him in writing prescriptions.

Dr. C. P. Lusk then Lecturer in Pharmacy was associated with the author in the preparation of the first edition and in spite of the fact that the book has been almost completely rewritten many traces of his practical knowledge and scholarship are to be found in the present edition. The stress of work which necessitated his retirement from the Department has led to his not taking the same part in the preparation of the second edition, that he did in the first. The author wishes to thank him for his great kindness in reading the copy of this edition and for his valuable criticisms and suggestions and regrets that Dr. Lusk felt constrained to withdraw his name from the title page.

The following works have been frequently consulted:— The British Pharmacopœia; The British Pharmaceutical Codex; Squire's Companion to the British Pharmacopœia; Ruddiman, Incompatibilities in Prescriptions; Elborne, The Elements of Practical Pharmacy and Dispensing; The Art of Dispensing, (published by the Chemist and Druggist); Fantus, Prescription Writing and Pharmacy; Bennett, Medical and Pharmaceutical Latin; The United States Pharmacopœia.

University of Toronto,  V. E. HENDERSON.
er 23 1911.

## CONTENTS.

Chapter I.    Definitions, Weights and Measures.
Chapter II.   The Methods and Definitions of Galenical Pharmacy.
Chapter III.  Posology.
Chapter IV.   Incompatibility.
Chapter V.    The Official Materia Medica.
Chapter VI.   The Non-official Materia Medica.
Chapter VII.  Notes on Prescribing.
Chapter VIII. Notes on Prescription-Writing.
Chapter IX.   Notes on Magistral Pharmacy.

# CHAPTER I.

## DEFINITIONS, WEIGHTS, AND MEASURES.

For the purpose of curing disease the medical practitioner makes use of many substances of animal, vegetable and mineral origin, as well as an increasing number of substances prepared by the chemist synthetically. The substances that are so used are known inclusively and collectively as the "MATERIA MEDICA." Any substance administered to a patient for the purpose of curing or alleviating disease may be termed a "DRUG." But not all substances that have been used by man as medicines are still in common use in civilized lands to-day, and many of the newer remedies, though highly lauded by their discoverers have not, and in many cases will not prove to be of sufficient merit to come into common favour. In consequence of this and as a guide to the physician and especially as an aid to his allies the pharmacists, most modern governments have caused to be prepared and published, books known as "PHARMACOPŒIAS." Such a pharmacopœia contains the correct legal or "*official*" names both in Latin and in the vulgar tongue of such substances of the materia medica as are judged by those who compile the pharmacopœia to be in common use in the country and to be of value to the physician. Further for the guidance of those who purchase crude drugs and prepare them for the patient's use the pharmacopœia contains accurate descriptions of the physical and chemical characteristics of the drugs and of the methods by which they are prepared for administration. The term "*official*" may be applied only to such drugs, preparations, methods and doses as are included in the British Pharmacopœia. This term must be carefully distinguished from the more inclusive term "*officinal*" which may be applied to any drug, etc., whether included in the Pharmacopœia or not, so long as it is in common use.

"PHARMACOGNOSY" is the science of the source and characteristics of the substances of the materia medica. This includes a knowledge of the natural history of all the plant, animal, and mineral products in the materia medica, as well as a knowledge of the methods of chemical preparation of those drugs that are produced synthetically, and a knowledge of the chemical and physical characteristics of all drugs. Some of the more important facts of the pharmacognosy of the drugs reviewed in this book will be referred to when they are individually considered.

"PHARMACY" is the art of the proper preparation of the substances of the materia medica for use (exhibition) and administration as medicines. This science may be divided into three branches. Firstly, *Chemical*

*Pharmacy*, or the preparation of substances of definite chemical composition, such as salts, acids, alkaloids, etc. This branch has now passed entirely out of the hands of the practising physician and almost entirely out of those of the practising pharmacist. Secondly, *Galenical Pharmacy*, or the preparation for administration in the form of medicine of drugs of indefinite chemical composition, which are, as a rule, products of plant or animal life and usually intimate mixtures of many chemical substances. Galenical pharmacy has now been almost entirely abandoned by the physician and only some of the simplest procedures are now carried out by him. The practising pharmacist as a rule no longer carries out the more complex galenical procedures but purchases many of his stock of galenicals from the larger pharmaceutical houses. Thirdly, *Dispensing, Magistral Pharmacy*, or the preparing and putting up in suitable form for the patient the drugs or their galenical preparations ordered by the physician.

"POSOLOGY" is the branch of medical science that deals with the doses of drugs and their preparations. The knowledge of this subject is of the utmost importance for the physician.

"PHARMACOLOGY" is the science that deals with the action of drugs upon the animal body. This science is often termed "PHARMACODYNAMICS"; the term "Pharmacology" being then used in a broader sense to include pharmacy, pharmacology, pharmacognosy, and posology. "THERAPEUTICS" is the art of applying the knowledge of these four sciences to the treatment of disease.

The Pharmacopœia also prescribes the systems of weights and measures, which are to be used in the operations of pharmacy. The older system, the IMPERIAL SYSTEM, is still almost exclusively used for Magistral Pharmacy, in spite of its obvious disadvantages, and in it alone the doses are given in the Pharmacopœia Britannica.

## MEASURES OF MASS OF THE IMPERIAL SYSTEM.

|  | Metric Equivalent. |  |
|---|---|---|
| 1 grain abbreviated gr | 64.7987 | mgms. |
| 437.5 grs.—1 ounce, abbreviated Oz. or ℥ | 28.349 | gms. ✓ |
| 7,000 grs. 16 oz. 1 pound, abbreviated lb | 453.59 | gms. |

Very commonly a weight known as a drachm (dr. or ℨ) equivalent to 60 grains (3.8879 gms.) is employed in prescribing and dispensing and more rarely the scruple ℈ equivalent to 20 grains. Both these weights are survivals of the Troy system and it is the common practice of pharmacists in spite of the ruling of the Pharmacopœia to use in dispensing the Troy ounce of 480 grains, unless there is some indication that the Imperial ounce is intended. (If the sign ℥ be used it is customary to dispense 480 grains, while if the word ounce be written the Imperial ounce would be dispensed).

## Measures of Capacity of the Imperial System.

|  | Metric Equivalent. |
|---|---|
| 1 minim abbreviated min. or m............................ | 0.0592 c.c. ✓ |
| 60 min. 1 fluid drachm abbreviated fl. dr. or fl. ʒ (often ʒ).. | 3.5515 c.c. |
| 8 fl. dr. 1 fluid ounce abbreviated fl.oz. or fl ʒ (often ʒ) .... | 28.4123 c.c. |
| 20 fl. oz. 1 pint abbreviated pt. or O..................... | 568.2454 c.c. |
| 8 pints 1 gallon abbreviated gal. or C..................... | 4.5459 L. |

## Relation of Volume to mass in the Imperial System.

1 minin is the volume at 62°F. of 0.9114 grains of distilled water.

1 fluid drachm is the volume at 62° F. of 54.6875 grains of distilled water.

1 fluid ounce is the volume at 62° F. of 1 oz. or 437.5 grains of distilled ✓ water.

1 pint is the volume at 62°F. of 1.25 ℔. or 8,750 grains of distilled water.

1 gallon is the volume at 62° F. of 10 ℔ or 70,000 grains of distilled water.

109.71 min. (approximately 110 min.) is the volume at 62 F. of 100 grains of water.

The more modern and convenient system adopted as an alternative system in galenical operations by the Pharmacopœia and as the preferred system in the British Pharmaceutical Codex, and in the Pharmacopœia of the United States of America, is the Metric System. It is the sole system in use throughout Europe, and will doubtless be adopted as the preferred system in the next revision of the British Pharmacopœia.

## Metric System Measures of Mass.

|  | Imperial Equivalent. |
|---|---|
| 1 milligramme abbreviated mgm. 0.001 gramme........... | 0.015 gr. |
| 1 centigramme abbreviated cgm. 0.01 gramme............ | 0.154 gr. |
| 1 decigramme abbreviated dgm. 0.1 gramme............. | 1.543 grs. |
| 1 gramme (The weight of 1 Millilitre of distilled water at 4°C.) abbreviated gm.........1 c.c............................ | 15.432 grs. ✓ |
| 1 decagramme abbreviated dkgm. 10 grammes..153.4 grs .. | 0.3527 oz. |
| 1 hectogramme abbreviated hgm. 100 grammes........... | 3.5274 oz. |
| 1 kilogramme abbreviated kgm, or kilo, 1000 grammes, 2.2046℔., | 35.27 oz. |

## Measurers of Capacity.

1 millilitre usually spoken of as a cubic centimetre and consequently abbreviated c.c. 0.001 litre.............. 16.89 min.
1 centilitre abbreviated CL 0.01 litre.................... 0.352 fl.oz.
1 decilitre abbreviated DL 0.1 litre..................... 0.1759 pint
1 litre (the volume at 4°C of 1,000 grammes of distilled water), abbreviated L..........................35.196 fl. oz., 1.7598 pints

The cubic centimetre that is a cube each of whose sides is a square centimetre is the unit of cubic capacity; it is usually considered to be of such a volume as to contain exactly one millilitre of distilled water at 4°C. It is according to the Pharmacopœia equivalent to 0.99984 millilitre. The term cubic centimetre is, however, used in place of millilitre throughout the Pharmacopœia and this book. The British Pharmaceutical Codex has introduced a new term intended to supersede this use of cubic centimetre The term introduced is "mil" an obvious abbreviation of millilitre. It has the added advantage that using the plan of the metric system, diminutives may readily be constructed to express quantities smaller than one cubic centimetre; ; this using this term 0.12 c.c. may be read twelve centimils, or 0.7 c.c. seven decimils.

*Domestic Measures.* A teaspoonful is a convenient but inaccurate measure and is considered as roughly equivalent to 1 fluid drachm (or 3.5 c.c.) a dessertspoonful is similarly considered to be equal to 2 fluid drachms (7c.c.) and a tablespoonful equivalent to a half fluid ounce (or 14 c.c.). A wineglassful though, too inaccurate for use in medicine is usually stated to be equal to 1½-2 fluid ounces, similarly a teacupful is estimated as 5 fluid ounces and a tumblerful as one half pint or 10 fluid ounces. A minim is considered to be equal to one drop but as the size of a drop varies with the viscosity of the fluid and the point from which it is dropped it is not to be considered an at all accurate measure. Graduated measures may now be obtained so cheaply that every physician should insist upon their use.

# CHAPTER II.

## THE METHODS AND DEFINITIONS OF GALENICAL PHARMACY

Drugs may be broadly classified as of (1) inorganic origin, (2) organic origin. They may also be divided into two classes, (1) pure chemicals, (2) galenicals. The pure chemicals are now prepared by neither pharmacist nor physician and in consequence the latter ordinarily needs to know nothing more about the methods of their preparation than what he has acquired as a student of chemistry. In regard to the methods of galenical pharmacy he must be better informed, as a knowledge of some of the methods and terms are essential in order that he may write prescriptions intelligently. In consequence some of the terms and methods are defined below.

*Fixed oils.*—(e.g. Castor Oil,*Olive Oil),—fluid esters of the higher fatty acids with glycerol (glycerin $C_3H_5(OH)_2$) obtained by expression from fruits, seeds, etc. They cannot be distilled without decomposition. They are freely soluble in ether, chloroform, carbon bi-sulphide, and benzene, slightly soluble in alcohol, but insoluble in water.

*Fats*, (e.g. Lard)—are solid esters of higher fatty acids and glycerol and are soluble in the same reagents as the oils. They are usually mixtures.

*Waxes*, (e.g. cera flava)—are usually mixtures of higher fatty acids and glycerol and higher alcohols.

*Volatile or Essential Oils*, (e.g. Oil of Cloves, Turpentine)—are usually mixtures of hydro-carbons chiefly fluids *terpenes* associated with more highly oxidized members *stearoptenes* which may be obtained in a solid state, (e.g. Camphor). They are usually isolated from plants by distillation. They are all soluble in ether, chloroform, carbon bi-sulphide, and benzene, fairly soluble in alcohol, slightly in water.

*Resins*, (e.g. Scammony Resin)—solid preparations obtained from oils by oxidation. The pharmacopœial resins are usually mixtures of resins as defined above and other bodies many of which are weakly acid. They are insoluble in water but soluble in alkaline solutions, alcohol, and ether.

*Oleo-resins*, (e.g. Copaiba) natural mixtures of volatile oils and resins semi-liquid in consistency.

*Balsams*, (e.g. Benzoin, Balsam of Tolu) resins or oleo-resins either liquid or solid which contain benzoic or cinnamic acids or both.

*Gums*, (e.g. Acacia and Tragacanth) solid or semi-solid exudations of plants which dissolve either partially or completely in water, forming a mucilage or an adhesive jelly, and are precipitated by alcohol. They are complex hydro-carbons yielding pentoses on hydrolysis.

*Gum-resins*, (e.g. Myrrh) mixtures of gums and resins.

*N.B. Students are strongly advised, when reading over this chapter, to look up in Chap. V the examples cited.

*Glucosides*, (the important pharmacopœial examples are,— Digitalin, Salicin, Santonin) active principles which may be readily broken up by acids or alkalies in the presence of water setting free glucose.

*Alkaloids*, (e.g. Morphine and Strychnine) nitrogenous organic bases usually pyridine derivatives which are generally crystalline though some are liquid. They are usually sparingly soluble in water, but readily in alcohol, chloroform, benzene, and ether. Like alkalies they form salts with acids. Those with inorganic acids are usually soluble in water, those with organic acids much less so.

*Tannins or Tannic Acids*,—These are weak acids containing a benzene ring, astringent in taste, freely soluble in alcohol and water. They occur very commonly in barks and roots and hence in pharmaceutical preparations of these they give precipitates with iron salts and some alkaloids. Their presence must be remembered when such preparations are prescribed.

In the following paragraphs the methods of preparation employed in galenical pharmacy are defined.

*Solution*,—the physico-chemical process by which a solid or fluid (the solute) disappears in a liquid (the solvent). The solute can usually be re-obtained chemically unchanged by any process which will remove the solvent.

There are a few solutions in the pharmacopœia in which a solid undergoes a chemical change by the action of the solvent during the process of solution, e.g. iron wire is dissolved in dilute acid to produce the Solution of the Perchloride of Iron.

*Extraction*,—the process by which a solvent (or menstruum) removes from a drug one or more of its soluble constituents. Four types of extraction are made use of by the pharmacist.

(i) *Infusion*—In this process a suitably finally divided drug is treated with either hot or cold water for a certain length of time, after which the fluid portion is strained off and retained and the solid portion rejected.

(ii) *Decoction.*—In this process the active principle is extracted by boiling in water.

(iii) *Maceration.*—In this process the drug is placed in a vessel, the solvent poured upon it, and left to stand for a suitable length of time with occasional agitation. The fluid is then filtered off; the marc or solid portion pressed out, the fluid thus obtained being added to the filtrate and the marc rejected.

(iv) *Percolation.*—In this process the drug is packed in a conical vessel (a percolator) with a small outlet at its lower end and moistened with the solvent which is added from time to time, and allowed to run off slowly from the lower outlet until a certain quantity of solvent has passed through. The marc is usually pressed out and the fluid obtained added to the percolate.

*Expression.*—In this process the drug is subjected to pressure and thus its juices are obtained.

*Filtration.*—In this process solids are separated from fluids by allowing the latter to pass out through a porous diaphragm.

*Dessication.*—In this process the watery constituents of drugs are got rid of by the aid of currents of either hot or cold air.

*Distillation.*—In this process volatile substances are separated from non-volatile or less volatile by the aid of heat. The volatile substances are passed over a cooled surface on which they condense and are collected.

*Pulverization.* By this process the drug is reduced to a very finely divided condition (or powder). The degree of fineness is determined by the number of meshes to the linear inch of the finest sieve through which the powder can pass. The sieves used contain 20, 40, 60, 80, 100 meshes to an inch. The simplest method of pulverizing the drug is by means of a mortar and pestle but in large pharmaceutical houses this end is usually obtained by means of a mill.

*Trituration.*—This term may be used as synonymous with pulverization, but more commonly refers to an intimate mixing and powdering of two drugs by means of a mortar and pestle or of a spatula.

## Galenical Preparations.

The crude drugs are rarely suitable for administration to the patient and in consequence are prepared by the pharmacist in various ways before being dispensed. The methods by which they are prepared have been defined above, but the forms in which they are dispensed are known by various names descriptive of the form in which they are dispensed or of the methods and solvents by which they are prepared and which must be known by the physician. The less important terms are printed in italics.

### Official Preparations.

*Acetum*, (Vinegar, e.g. Acetum Scillæ)—A solution of active principle made by solution or maceration with acetic acid.

**Aqua**, (Water)—A solution of volatile substances in water,—some are simple solutions, such as Aqua Chloroformi, while others (Aqua Cinnamomi) are prepared by distillation if made in accordance with the British Pharmacœpia. In the colonies a pharmacist is allowed to make a Water containing a volatile oil by first triturating the oil with calcium phosphate and then suspending the triturate in water. The oil being very finely divided remains suspended in the water. The Waters are very commonly used in mixtures as flavoring vehicles for the administration of less pleasant drugs. They are in some cases of value on account of their own pharmacological action.

*Charta*, (Paper, e.g. Charta Sinapis)—A strip of cartridge paper smeared with a preparation of an active drug. They are applied to the surface of the body.

*Collodium*, (Collodion, e.g. Collodium Flexile).—A solution of Pyroxylin in Ether and Alcohol, either alone or containing also some other drug in solution. It is applied to the skin and when the ether or alcohol has evaporated it leaves a thin film upon the surface.

*Confectio,* (Confection, e.g. Confectio Sennæ)—A soft sticky mixture of sugar or syrup with some active drug. These preparations are very little used at the present day.

*Decoctum,* (Decoction, e.g. Decoctum Haematoxyli)—There are only three official decoctions and it is rarely that they are used .

**Emplastrum,** (Plaster, e.g., Emplastrum Plumbi)—A preparation composed of some active drug incorporated with an adhesive and permanent base such as lead oleate, soap, or resin, and of such consistence that they can be spread upon linen, muslin, or leather and will remain adherent if applied to the skin.

**Extractum,** (Extract)—An extract which has been evaporated to a solid or semi-solid consistence. If the extract has been made with water it is known as an *Aqueous Extract,* if with alcohol, as an *Alcoholic Extract.* As in proportion to their bulk they contain 2-6 times as much of the active constituents as the crude drug, they have a much smaller dose and are of especial value in the preparation of pills.

**Extractum Liquidum,** (Liquid or Fluid Extract, e.g., Extractum Filieis Liquidum.)—A concentrated extract of such strength that each fluid ounce of the product of the extract represents the active principles of an ounce by weight of the drug. They may be made with alcohol, water, or ether. They are very convenient preparations for incorporating in a mixture, but as they may contain substances which are soluble in the solvent with which they are made, but not soluble in other solvents they may give rise, if mixed with other solvents, to precipitates of inactive ingredients which may be filtered off.

*Extractum Viride,* (Green Extract, e.g. Extractum Hyoscyami Viride)— These are the partially or completely dried juices of plants obtained by expression. As they contain the active constituents in a concentrated form and are usually adhesive, especially if a little water is added, they may frequently be used for the preparation of pills.

*Glycerinum,* (Glycerin, e.g. Glycerinum Tragacanthæ).—A solution of a drug in glycerin. They are useful on account of the special solvent or preservative properties of glycerin.

**Infusum,** (Infusion, e.g. Infusum Digitalis)—an extract prepared by infusion.

*Injectio Hypodermica,* (Hypodermic Injection, e.g. Injectio Morphinæ Hypodermica)—A solution of a potent drug in water. They are administered by means of a syringe underneath the skin. Great care must be taken that they are sterile. This is obtained by boiling or by care in preparation or by the addition of an antiseptic.

*Lamella*, (Disc, e.g. Lamella Atropinæ)—a thin transparent plate of gelatin and glycerin containing a small quantity of an alkaloid. It is placed in the conjunctival sac and allowed to dissolve. In this way a purely local effect of the drug stuff may be obtained.

*Linimentum*, (Liniment, e.g. Linimentum Camphoræ)—A liquid preparation dissolved in a menstrum of alcohol, water, or oil, with soap, camphor, or glycerin. They are intended to produce a local action of the drugs they contain by being rubbed into the skin.

**Liquor**, (Liquor or Solution, e.g. Liquor Arsenicalis)—Solutions of definite chemical substances in water. They are suitable preparations for dispensing in mixtures.

*Liquor Concentratus*, (Concentrated Solution, e.g. Liquor Calumbæ Concentratus)—Weak extracts of such a strength that two parts represent one of the crude drug. They are intended to facilitate the making of infusions, for which purpose they need only be diluted with water.

*Lotio*, (Lotion, e.g. Lotio Hydrargyri Flava)—A suspension of a drug in water. It is applied to the skin as a wash or in lint saturated with it.

*Mel*, (Honey, e.g. Mel. Boracis)—Syrupy liquids containing honey.

**Mistura,** (Mixture, e.g. Mistura Ferri)—A preparation containing drugs dissolved or suspended in water.

**Mucilago**, (Mucilage, e.g. Mucilago Acaciæ)—A viscid solution of gums or starch. Used for the making of pills or the suspension of insoluble powders in mixtures.

*Oxymel*, (Acidulated Honey, e.g. Oxymel Scillæ)—A preparation containing honey and acetic acid.

**Pilula**, (Pill, e.g. Pilula Ferri). A spherical or spheroidal mass which contains one or more potent drugs held together by some adhesive substance known as the excipient. In this way a disagreeable drug-stuff of small bulk may readily be administered to a patient.

**Pulvis**, (Powder, e.g. Pulvis Ipecacuanhæ Composita)—A mixture of drugs reduced to a fine powder. When two or more unimportant drugs are included in a powder it is usually known as a compound powder, Pulvis Composita. Usually only insoluble drug-stuffs are administered in this form. They are given by the mouth.

**Spiritus,** (Spirit, e.g. Spiritus Camphoræ)—Solutions of volatile substances prepared by either simple solution in Rectified Spirit or by distillation. They are frequently ingredients of mixtures.

*Succus*, (Juice, e.g. Succus Conii—)The juices of fresh plants obtained by expresssion and preserved by the addition of Alcohol.  Unimportant.

**Suppositorium,** (Suppository, e.g. Suppositoria Morphinæ)—Conical masses usually made by incorporating some drug with Oil of Theobromal (This is a solid). They are made to weigh about fifteen grains each and are used by inserting them into the rectum. Suppositories made for use in the vagina are made to weigh about a drachm and are called pessaries, while those used for the urethra are elongated rods, made with Cocoa Butter or gelatin, and are called Bougies.

**Syrups,** (Syrup, e.g. Syrupus Tolutanus)—Viscid liquids prepared by dissolving active medicines in a syrup made from cane-sugar and water. Used as flavouring vehicles and to suspend insoluble powders in mixtures.

*Tabella,* (Tablet)—The only official tablet is that of Nitro-glycerin which is composed of nitroglycerin incorporated with Chocolate and moulded to a flat, circular shape.

**Tinctura,** (Tincture)—Fluid preparations of drugs prepared by solution, maceration or percolation with alcohol. That of Lobelia alone is prepared with ether. Those containing more than one active principle are known as *Compound Tinctures*. They are weaker in pharmacological action than the Liquid Extracts and are the most suitable form in which drugs soluble in alcohol may be incorporated in mixtures which contain alcohol as the main solvent. (If much water is present ingredients not of pharmacological importance which are present in the tincture may be precipitated).

*Standardized Tinctures* are such as must show by assay a certain quantity of certain of its constituents. The term standardized may also be applied to Extracts.

*Trochiscus*, (Lozenge, e.g. Trochisci Potassii Chloratis)—A large dry tablet prepared by mixing an active drug or drugs with Refined Sugar and Powdered Acacia forming a mass by the aid of one of four bases, Fruit Basis (Black Current Paste), Rose Basis (Rose Water), Tolu Basis (Tincture of Tolu) or Simple Basis (Water) and then dividing the resulting mass with a suitable mould into lozenges of definite weight, which are then dried.

**Unguentum,** (Ointment, e.g. Unguentum Hydrargyri)—A preparation made by incorporating solutions or finely divided drugs with a fatty base. This is commonly Wool-fat, Lard, or a Paraffin. They are smeared or rubbed into the skin.

*Vinum*, (Wine, e.g. Vinum Ipecacuanhæ)—Solutions of drugs in either Sherry (Vinum Xericum) or Orange Wine (Vinum Aurantium). They are but little used.

**Non-Official Preparations.**

**Cachet,** (Cachet or Konseal).—A cachet is made of two concave plates of rice paper within which the medicament is enclosed and which is then sealed by moistening the contiguous borders of the plates with water. They offer an elegant method for completely covering nauseous and insoluble powders which are too bulky to be made into pills.

**Capsula,** (Gelatine Capsule).—Capsules are made in hard and soft varieties. The first are hollow receptacles, covered by a lid made of the same shape and accurately closing it, and composed of gelatine, acacia and sugar. The soft variety made by substiting Glycerin for the sugar, are ovoid in shape and are closed, after being filled, simply by placing a drop of the gelatine solution over the open end. The substances introduced may be bulky powders, semi-solid pill masses, and such fluids as will not dissolve the gelatine, as the Oils. Watery solutions may be administered by this means if given immediately but this method is not recommended.

**Cataplasma,** (Poultice).—A poultice is a means of applying moist heat to the surface of the body. It may be made of Linseed Meal, Bran, or any other bland substance capable of retaining heat and moisture. Sometimes they contain also more active substances such as Mustard, small quantities of Laudanum, or some of the antiseptics as Boric Acid and the Volatile Oils. In the latter case a base of Kaolin is used as in the Cataplasma Kaolini of the United States Pharmacopœia.

*Ceratum,*(Cerate).—These are fatty mixtures made as ointments but containing wax which gives then a firmer consistence. They are therefore valuable as local applications.

*Collyrium,* (Eye-wash).—A solution of a drug or drugs dissolved in water. They are dropped into the conjunctiva.

*Elixir,* (Elixir).—A solution of active remedies in a mixture of Syrup and Alcohol which has been made aromatic by the addition of some of the Essential Oils. As a class they are related to both the Tinctures and the Spirits but are usually of feeble strength. Some are used simply as flavourings or as vehicles for less pleasant drugs.

**Emulsio,** (Emulsion).—A mixture of oil and water in which the oil is suspended by the use of a mucilage or in which it has been partially saponified by the action of an alkali.

**Enema,** (Enema or Clyster).—A liquid preparation for injection into the rectum. These may be medicated or nutrient in character. In the first any drug capable of acting upon the mucous membrane of the rectum or which can be absorbed, and thus permitted to exercise its general effect may be used. The latter usually consist of easily absorbed food such as pre-digested milk or eggs.

*Fumigatio* (Fumigation).—Fumigation is the act of subjecting the body or any object to the action of fumes or vapors, as in the burning of sulphur for its disinfecting properties, or as in the fumigation of calomel in the treatment of Syphilis.

**Serum,** (Serum).—The purified serum obtained under the most rigid aseptic precautions from animals which have been inoculated with living bacteria or their products. The Antidiphtheritic Serum is the best known and understood.

*Tabella*, (Tablet).—Tablets are of three kinds (1) those made by compression called Compressed Tablets and for the making of which the drug used must be in the form of a granular powder, and which may be coated with sugar or gelatine, if desired; (2) those made by moulding without compression, for which drugs of small bulk are essential, incorporated with milk-sugar as a base, and which are not coated, called Tablet Triturates; and Hypodermic Tablet Triturates which are prepared from potent drugs under aseptic precautions with a base of Milk-sugar or better of Granulated Sodium Sulphate.

The official Chocolate tablet will be found grouped with the official preparations.

*Tampon.*—Is a plug of medicated absorbent Cotton or Lamb's Wool used in a natural or in an artificial cavity of the body for the purpose of arresting hæmorrhage or for correcting the secretions.

# CHAPTER III.
## POSOLOGY.

The British Pharmacopœia makes the following statement in regard to the doses as given in it. "The doses mentioned in the pharmacopœia are intended to represent the average range in ordinary cases, for adults. They are meant for general guidance, but are not authoratively enjoined by the council. The medical practitioner must act upon his own responsibility as to the doses of any therapeutic agents he may administer." This statement is a very important one and one that should be thoroughly understood by every medical practitioner. Firstly, the official doses represent the average range in ordinary cases. The deviations from the ordinary that are most likely to be met with must be considered. *Weight.* Roughly the larger and more robust the individual the larger the dose of most drugs that my be given to him. Small and weakly individuals should always receive small doses of any remedy at first. *Sex.* Women are often said to be less resistant to the action of drugs than men are but as a rule little distinction is made between the sexes. It must however be born in mind that at the time of pregnancy or menstruation any drugs that bring about changes in the blood-supply to the uterus or that would set up movements in its musculature should be either entirely avoided or given in very small doses and with caution. Also it must be remembered that many drugs are excreted in the milk and may readily make the milk unpalatable or even dangerous to a suckling child. Amongst the drugs excreted by the mammary glands are the oils of anise and dill, turpentine, copaiba, the purgative principles of rhubarb, senna, and castor oil, opium, iodine, also some of the metals antimony, arsenic, iron, lead, mercury and zinc. *Idiosyncrasy.* Every person differs from all others more or less. Each person is not only physically but also chemically a distinct individual. These personal differences are usually quantitively so small as to occasion little or no difficulty but occasional individuals are met with who deviate very widely from the normal in respect to some one or more drugs. Such individuals as are abnormally affected by any drug are said to have an idiosyncrasy for the drug. Drugs in regard to which idiosyncrasy is likely to be encountered are morphine, and its allies, mercury, bromides, copaiba, arsenic, iodides, quinine, etc. Idiosyncrasy is often an inherited characteristic. *Tolerance.* The continued use of a drug is very apt to make any individual less susceptible to its pharmacological action and to necessitate the administration of larger doses, this is known as tolerance. Tolerance often occurs with alcohol, morphine, arsenic, vegetable purgatives. cocaine.

*Increased susceptibility* to the action of the drug due to its continued administration also occurs. It rarely gives trouble except with those drugs such as Digitalis which can be more readily absorbed by the normal body than they can be excreted by it. *Disease* may readily influence the absorption of a drug on the one hand or interfere with its excretion on the other. For example a large skin-wound may readily absorb a poisonous dose of carbolic acid or iodoform. Or increased acidity in the stomach may lead to a larger absorption of bismuth salts than is normally the case. Diminished excretion by the kidney will lead to a more prolonged action of strychnine.

Secondly, the doses of the pharmacopœia are doses for adults. For children much smaller doses must be given. The rule suggested by Young is perhaps the best for calculating the dose for a child. Multiply the adult dose by the age of the child and divide by the age of the child plus 12. Thus for a child of three, the dose would be $\frac{3}{3+12}$ or 1/5th; for an adult dose of 15 min. it would be $\frac{15 \times 3}{3+12}$ or 3 min. Another rule suggested by Brunton is to multiply the age at the next birthday by the dose and divide by 25 (the assumed adult age), or perhaps better multiply the dose by four times the age at the next birthday and divided by 100: for the example stated above that would be $\frac{4 \times 4 \times 15}{100}$ or 2.4 min. roughly 2½ min. Young children are particularly prone to be affected by morphine and its allied drugs, but are proportionately little influenced by strychnine, and alcohol.

Persons above the age of sixty are proportionately more affected by drugs than are younger persons, so that by adults must be understood persons between 20-60 years of age. Persons over 60 should receive roughly ¾ and persons over 85 roughly ½ of the adult dose, save in the case of purgatives to which the aged are often very refractory.

Thirdly, the frequency of repetition makes a great difference in the size of dose to be administered. The more frequently the drug is to be administered the smaller the dose should be.

Fourthly, the time of day makes as a rule but little difference, except with the case of drugs meant to bring on or increase a normal daily condition. For example a larger dose of a hypnotic such as chloral would be necessary to produce sleep during the day than at night. Also purgatives can best be given at such an hour that they will take effect at the hour of the patient's daily defecation. For this purpose calomel and aloes must be given some eight hours in advance, while purgative salts act within an hour or so.

The presence or absence of food in the stomach makes a great difference in the rapidity with which drugs are absorbed and in the quantity coming in contact with the wall of the stomach and so irritating it, and as a consequence of this larger quantities of any drug irritant to the stomach may be given immediately after than before meals.

Fifthly, *Synergists* are drugs having the same pharmacological final effects though the manner of action may be slightly different. For example, colocynth, aloes and potassium sulphate are synergists, as they are all purgatives. All of these occur in the Compound Pill of Colocynth. It is often an advantage, and this is especially the case in the administration of purgatives to include in a prescription two or more synergists. As in the instance mentioned above, when synergists are administered together it is necessary to give any one drug in only a fraction of its full dose.

Finally, the pharmacopœial doses are not enjoined and the practitioner must use his own judgment. In many cases it is quite allowable to exceed the pharmacopœial dose if the effects wished for are not achieved by its administration and the physician should carefully watch each and every patient and convince himself that the drugs given are really producing the wished for action. In other words he must not take it for granted that because he gives a pharmacopœial dose of any drug that he must as a consequence get the described pharmacological action.

The doses of the pharmacopœia are usually, and unless otherwise is stated, for administration by the mouth. Many drugs can however with advantage be administered by hypodermic, intramuscular or even by intravenous injection. Owing to the more rapid absorption as a rule of drugs given by these methods and to the certainty that they will be absorbed in their entirety it is not necessary that such large doses be given. In those cases in which drugs have given by intravenous injection only a small fraction of the dose given by the mouth was used. For drugs given subcutaneously (hypodermically) about one half of the dose is given that would be used if given by the mouth. Drugs given by inunction must be given in larger doses than would be used if they were given by the mouth. The same is true as a rule for drugs given by the rectum if they are intended to have a general action.

# CHAPTER IV.

## INCOMPATIBILITY.

Many of the drugs and preparations of the materia medica may be and are given alone, but many others only in combination. There are a few which are never or almost never given in combination with other drugs, but the majority of drugs and their preparations are at times given in combinations, which are often very complex. The selection of drugs and preparations to be used in combination with each other requires a great deal of care to avoid unwished for changes being brought about by thier admixture. Two drugs are said to be "INCOMPATIBLE," when on being brought into intimate contact with each other unwished for changes either physical or chemical are brought about or when their pharmacological actions would so interfere with each other as to be detrimental. It is by no means an infrequent occurrence for a physician to prescribe together two medicines which have almost opposite pharmacological actions but he does so in such proportions that the action of the one serves but to correct some undesired action of the other.

Incompatibility dependent upon the differing pharmacological actions of the drugs administered together is known as THERAPEUTICAL or better PHARMACOLOGICAL INCOMPATIBILITY. An extreme example would be the administration of atropine and pilocarpine together.

Incompatibility dependent upon chemical and physical changes can only occur when the drugs are brought into intimate physical contact either by trituration in a mortar (the cases in which incompatibility is apt to make be made manifest in this way will be found mentioned in paragraphs IVd. and V.) or by solution. The incompatibility due to chemical changes occuring between preparations dispensed together is known as CHEMICAL INCOMPATIBILITY. The changes may be of several types and may be classified as follows:—

1. Resulting in chemical change without any visible change.

(*a*) The neutralization of *acids by bases.**

(*b*) The breaking up of *glucosides by acids* (sugar is set free and the glucoside loses in activity).

(*c*) The action of acids on the activity of pancreatic ferments and of alkali on gastric ferments.

2. Resulting in precipitation of newly formed chemical substances due to the interaction of two other chemical substances in solution.

*Important cases are printed in italics.

(*a*) Salts of the *alkaline earths* are precipitated by alkali hydroxides and *carbonates, phosphates*, borates, oxalates (the corresponding insoluble salts of the alkaline earths being formed). The free acids which would form corresponding salts are also incompatible.

(*b*) *Salts of the metals* in solution are incompatible *with* hydrates, *carbonates, phosphates*, oxalates and the corresponding acids; in many cases with proteins, *tannins, acacia* and often alkaloids and phenozone. Silver, *mercurous,* lead, and *bismuth* salts also *with bromides* and *iodides*: the *same metals and calcium,* barium and strontium, *with sulphates* and sulphuric acid.

(*c*) *Hydrates or carbonates of the alkalies,* sodium, potassium, and ammonia *with* salts of *metals* and *alkaline earths,* and with *alkaloids* and some *glucosides.*

(*d*) *Alkaloids* form insoluble salts with other organic acids than acetic and citric; the free alkaloid being very much less soluble than the salts is precipitated *by alkali hydrates* and *carbonates* and by *borax*. Ammonium carbonate and the bicarbonates do not so readily cause precipitation. *Iodides, bromides, salicylates, benzoates*, usually cause a precipitate *tannic acid*, and *iodine in a solution of mercuric iodide;* precipitation may be prevented in many of these cases by from 15-50% of alcohol. About 15% suffices to prevent that by bicarbonates and carbonates. Alkaloides may give a precipitate with many metallic salts especially those of mercury.

(*e*) Proteins are precipitated by alkaloids, many metal salts, tannin and alcohol.

3. Resulting in a change of colour owing to the formation of some soluble but undesired body owing to the interaction of two other substances in solution.

(*a*) Giving an objectional appearance *tannic* and gallic *acids and iron* preparations, ammonia and carbolic acid; gallic acid and thymol. *Ferric chloride with salicylates,* carbolic acid, creasote, guaiacol, salol, acetanilid, phenazone, phenacetin, oils of wintergreen, cloves, pimenta, and thyme, podophyllin, *aloin*, gamboge, asafetida, storax, myrrh, balsam of Peru, balsam of Tolu, *morphine* and apomorphine.

(*b*) The change in colour is the indication of a chemical change objectionable from the pharmacological side also. *Salicylates, phenozone, acetanilid, with* the free nitrous acid in *Spirits of Nitrous Ether* (isonitroso-compounds are formed).

4. Resulting in the chemical splitting of one of the bodies and the formation of an undesired body.

(*a*) Resulting in the freeing of a volatile body, which may in part or entirely, dependent upon the amount formed, remain in solution. Hydrochloric acid with nitric acid (nitrous oxides freed); strong acids with alcohol (ethers); *acids and carbonates;* acids and sulphides; *mineral acids with iodides, bromides, and chlorates; ammonium salts* and hydrates and *carbonates of the alkalies.*

(*b*) Resulting in the freeing of a liquid body, *chloral* and butylchloral *with alkalies* (chloroform freed).

(*c*) Resulting in the freeing of dextrose or other sugar, *glucosides with acids* and alkalies.

(*d*) Resulting in liberation of so much gas suddenly as to cause an explosion. Chromic acid, concentrated nitric acid, nitrates, permanganates, chlorates, with such substances as sulphur and sulphides, sulphites, iodides, phosphorus, hypophosphites, reduced iron, and many organic bodies, sugar, tannin, etc. These reactions only occur when the dry substances are triturated together or in some cases when mixed in very concentrated solutions.

5. In some cases when two solids are triturated together a soft sticky or a damp mass, or a liquid is formed: the reaction is probably always to a certain extent chemical. Such substances are camphor, carbolic acid, thymol, phenozone, phenacetin, chloral, sodium phosphate, lead acetate. Details will be found under the various drugs.

PHARMACEUTICAL OR PHYSICAL INCOMPATIBILITY.

1. Resulting in precipitation of one of the ingredients in solution owing to its decreased solubility when its solvent is diluted by another liquid. The dilution of aqueous solutions of acacia, proteins, salts (if strong). and emulsions with alcohol. Some gums as well as starches and dextrins are similarly precipitated by alcohol. In some of the cases that will occur under this rule the precipitate is not an important constituent, for example, the Liquid Extract of Cascara Sagrada gives a precipitate with alcohol, the precipitate consists however of unimportant constituents and may be filtered off. The dilution of alcoholic solutions of resins, oleoresins, oils, etc., by water. In some of these cases also the precipitate is unimportant, for example, Liquid Extract of Nux Vomica and water.

It must be distinctly understood that at times it is advisable or even necessary to order incompatibles in a prescription. Attention might be called to the fact that the Pharmacopœia contains such formulæ, for example the Lotio Hydryrgyri Nigra, and the Mistura Ferri Composita. Whenever the physician orders such a preparation he should warn the patient that the bottle will contain a deposit. It is only rarely that one should write such a prescription as will involve an uncorrected incompatibility. No prescription should ever be written which if dispensed would lead to the precipitation of any highly active ingredient, as in the cases of such a precipitate the patient might readily be poisoned by getting an over-dose of the potent precipitate in the last dose. The practitioner should make it the rule to send out preparations free from precipitate and of an attractive colour. In some cases the incompatibility may be overcome; for example the carbonates and the bicarbonates, the bromides and iodides of the alkaloids while less soluble in water than the usual salts, are comparatively soluble in alcohol, and hence the addition of alcohol will prevent the precipitation. In other cases it may be possible by increasing the viscosity of the mixture by the addition of acacia, tragacanth or syrup to prevent the formation of a precipitate or much more often the addition of one of these ingredients will so prevent the clotting of the precipitate that it may be safely dispensed with a "Shake the Bottle" label.

There are some substances such as, the salts of silver, phenacetin, phenazone, potassium iodide and calomel that react with so many other drugs that it is preferable to administer them alone, or in simple solutions with a flavouring reagent or in pills.

# CHAPTER V.

## THE OFFICIAL MATERIA MEDICA.

In this chapter only the drugs of the British Pharmacopœia are considered. For the convenience of the student and for the purpose of reference, the drugs have been arranged in alphabetical order, save that the preparations of any drug immediately follow it. As preparations of any drug are considered such galenicals as bear the name of the drug as an important part of their title, (under this rule Pilula Saponis Composita is classed as a preparation of Sapo.) or such galenicals as contain the drug as their important constituent, (under this rule Pipula Saponis Composita is classed also with Opium). That a galenical is considered as a preparation of a drug is indicated by its name being set further from the margin (indented) than is the name of the drug under which it is classed as a preparation. This rule is adhered to for galenicals but certain of the active principles, such for example as Atropine, though classed with the preparations of Belladonna, are treated in all other respects as separate drugs. There are also several galenicals especially among the Liquors, e.g. Liquor Trinitrini, whose active principle is not official, and these will be found classed according to their official name. The titles of galenicals are as a rule placed immediately after the preparation from which they are prepared and their title is again set further from the margin of the page. The salts of any base appear under the general heading of the base as though they were preparations of it.

Further for the advantage of the student the drugs have been divided into four classes,*indicating their relative importance. The names of drugs of the first class, those of pre-eminent importance, are printed thus **OPIUM:** these drugs the student must master thoroughly. The drugs of the second class have their names printed thus, **Acetanilidum:** these drugs should also be thoroughly studied. In the third group are included many useful and frequently used drugs, and with them the student should be familiar: their names are printed thus, ACACIA GUMMI. The drugs of the fourth class are of minor importance in use or in activity: their names are printed thus, *Ammoniacum*. Even important drugs have, however, preparations that vary amongst themselves in importance from the practical standpoint, and an attempt has been made to indicate this by placing before their titles a superior numeral, thus, [1]Tinctura Opii. The superior numeral one will indicate that the preparation is of importance, the numeral three will indicate that the pre-

---

*The author thoroughly realizes that any such classification of drugs and of preparations as has been adopted is open to criticism both in principle and in the details of the classification itself. Any such classification must be largely a personal one. It has, however, been resorted to it with the view of aiding the student.

paration is relatively of no importance: the numeral two indicates roughly a preparation of an importance mid-way between the other two. It will be noted that drugs of the fourth class have as a rule preparations without a numeral prefix In some cases none of the preparations of a drug seemed worthy of the numeral one or even of the numeral two, this indicates that the drug is much more important than any of its preparations.

The official Latin name of the drug is always given; its English equivalent only when difficulty might arise in translating the Latin or where other considerations seemed to render the giving of it an advantage. Important synonyms are in many case also given: they are always enclosed in brackets. The dose is given in both the Imperial and the Metric systems. The Imperial as being the official dose is given the preference. The doses in the Metric system are as a rule those of the British Pharmaceutical Codex, though the term "mil" and its diminutives in spite of its very obvious advantages have not been adopted. Doses enclosed in brackets are not official.

No attempt is made to give the full Pharmacopœial definition of any drug or description of its physical or chemical characters, enough only is given to draw the attention of the student to some of its outstanding characters a knowledge of which may be of advantage to him. The formulæ for the preparation of galenicals are as a rule taken from the British Pharmaceutical Codex. In this the Pharmacopœial formulæ have been recalculated so that the total of the quantities of the ingredients at the end of the process of preparation will aggregate 100. The advantages of this centesimal system are obvious, the principle one, being the ease with which the percentage strength of any ingredient may be seen. It is an added advantage in a laboratory where the dispensing is carried out in the Metric system. No attempt has been made to give a detailed description of the steps to be pursued in the preparation of any galenical, but only enough is indicated to aid the student to use them intelligently in dispensing and prescribing. For a knowledge of the steps in preparation of galenicals the student is referred to the Pharmacopœia or the Codex. Where the formula of any preparation is given the first quantity preceeded by a dash indicates the quantity of the drug or preparation under which the preparation in question is classed.

The more important solubilities of the drugs are also given and are stated for room-temperature, 15.5 C, unless the word "hot" is used as meaning boiling (by cold is meant 15.5 C as opposed to boiling). Solubilities are always expressed in parts by weight. By water, distilled water is always meant and by alcohol, 90% alcohol (Rectified Spirit).

The important incompatibilities are given with often an indication of the chemical change occuring. In some cases methods of overcoming or lessening the incompatibility are also given.

ACACIÆ GUMMI. GUM ACACIA.—The gum exuded by Acacia Senegal and other species. In rounded or ovoid brittle tears, either colourless or of a pale yellowish tinge, often opaque due to numerous small fissures. Small angular fragments may occur with glistening faces. Nearly inordorous, taste bland and mucilaginous. Insoluble in alcohol, soluble 1 in 1 of water forming a viscid, slightly acid solution.

Incompatibles, alcohol and sulphuric acid. Borax, ferric, and lead salts render it gelatinous.

[1]Mucilago Acaciæ.—40; Water, to100.

**Acetanilidum. Acetanilide.** (Antifebrin.) Dose, 1-3 grs; ½-2 dgms.

Phenylacetamide, $CH_3CONHC_6H_5$. Colourless, inodorous, glistening, lamellar crystals with a pungent taste. Soluble 1 in 200 of cold, 1 in 18 of hot water; 1 in 4 of alcohol; and in ether and chloroform.

Incompatibles, strong solutions of sodium and potassium hydrate (anilin formed); bromides and iodides, spirits of nitrous ether, amyl nitrite (diazo compounds formed); a red colour is given with tincture of the chloride of iron: forms liquids if triturated with phenol, resorcin, and thymol, and a damp powder with chloral.

**Aceta** (see Cantharis, Ipecacuanha, Squill).

ACIDUM ACETICUM. ACETIC ACID. A colourless, pungent liquid miscible with alcohol and water. Contains 33% by weight of real acid, $CH_3$ COOH. Sp. Gr. 1.044.

[1]Acidum Aceticum Dilutum.—Dose, ½-2 fl. dr.; 2-8 c.c. —12.47; Water, to 100. Contains 4.27% of real acid.

[2]Oxymel. Dose, 1-2 fl. dr.; 4-8 c.c. —10; clarified Honey, 80; Water to 100.

*Acidum Aceticum Glaciale.*—A colourless, pungent liquid or crystalline mass. Contains 99% by weight of real acid. Sp. Gr. 1.058.

Acidum Arseniosum (see Arsenium).

Acidum Benzoicum (see Benzoinum).

Acidum Boricum (see Boron).

**ACIDUM CARBOLICUM. PHENOL.** (Carbolic Acid). Dose 1-3 grs.; ½-2 dgms.

$C_6H_5$. OH. Small, colourless, deliquescent crystals; odour characteristic; taste sweet and pungent; with a caustic action on the skin and mucous membranes, turning them white. Soluble in alcohol, ether, chloroform glycerin, fats, oils and solutions of alkalies; liquified by 10% of water, forms a clear liquid with 30-40% of water, and completely dissolves in 12 parts of water.

Incompatibles, ferric and mercuric salts in solution, hydrogen peroxide and potassium permanganate, gelatin and albumin: forms liquids if triturated with chloral, acetanilide, camphor, phenazone, phenacetin, salol, menthol, thymol, resorcin, and naphthol.

[1]Acidum Carbolicum Liquifactum.—Dose, 1-3 min.; 0.06-0.2 c.c. —90; Water, 9.

[1]Glycerinum Acidi Carbolici.—20; Glycerin, to 100.

[1]Suppositoria Acidi Carbolici.—Each suppository contains 1 gr. —6.70; White Beeswax, 13.40; Oil of Theobroma, to 100.

[3]Trochiscus Acidi Carbolici.—Each lozenge contains 1 gr. Made with Tolu Basis.

[1]Unguentum Acidi Carbolici.—4; Glycerin, 12; Paraffin Ointment, 84.

*Acidum Chromicum. Chromic Anhydride (Chromic Acid.)*—$CrO_3$. Crimson, odourless crystals, deliquescent, caustic to skin and mucous membranes.

[2]Liquor Acidi Chromici. Solution of Chromic Acid.—An aqueous solution containing 25% $CrO_3$.

**Acidum Citricum. Citric Acid.** Dose, 5– 20 grs; 3-12 dgms. $C_3H_4$ OH. $(COOH)_3$. $H_2O$. Large, colourless crystals. Soluble 1 in ¾ of cold, 1 in ½ of hot water; 1 in 1 of alcohol. 17 gr. neutralizes 24 gr. $KHCO_3$,- 20 gr. $K_2CO_3$,- 20 gr. $NaHCO_3$,- 34 gr. $Na_2CO_3$,- 12 gr. $(NH_4)_2CO_3$- 11 gr. $MgCO_3$.

Citrates are incompatible with lead and silver salts in solution, and with quinine (quinine citrate is soluble 1 in 800 of water).

*Acidum Gallicum. Gallic Acid.*—Dose, 5-15 grs.; 3-10 dgms. $C_6H_2$ $(OH_3)$ COOH, $H_2O$. Acicular, slightly brownish crystals, odourless, faintly acid taste. Soluble 1 in 100 of water; 1 in 5 of alcohol.

Incompatibles, alkali hydrates, ammonia, lime water, lead and iron salts, oxidizing agents.

ACIDUM HYDROBROMICUM DILUTUM.—Dose, 15-60 min.; 1-4 cc.. A colourless, odourless liquid, containing 10% by weight of hydrogen bromide, Roughly 7 c.c. (2 dr.) contain as much bromine as 1 gm. (15 gr.) potassium bromide.

Incompatibles, alkalies and their carbonates, metallic oxides, silver and lead.

**Acidum Hydrochloricum.** A watery solution, containing 31.79% hydrogen chloride, HCl8 by weight. Sp. Gr. 1.16.

[1]Acidum Hydrochloricum Dilutum. Dose, 5-20 min.; 0.3-1.2. c.c. —30.18; Water, to 100. Contains 10.58% by weight of hydrogen chloride.

Incompatibles, alkalies and their carbonates, metallic oxides, salts of silver, lead, and antimony.

**Acidum Hydrocyanicum Dilutum.** Dose, 2-6 min.; 0.1-0.4 c.c. A colourless liquid with a characteristic odour, containing 2% of hydrogen cyanide, HCN, volatile and very poisonous.

Incompatibles, copper, iron and silver salts, mercuric oxide, sulphides, morphine.

*Acidum Lacticum. Lactic Acid.*—A colourless, odourless, slightly hydroscopic liquid, containing 75% of real acid $CH_3$ CHOH. COOH. Mixes freely with alcohol, water, and ether.

Incompatibles, albumin, most metallic salts in solution, nitric acid and potassium permanganate.

**Acidum Nitricum.** A clear, colourless liquid emitting corrosive fumes, containing 70% by weight of hydrogen nitrate, $HNO_3$, Sp. Gr. 1.42.

[1]Acidum Nitricum Dilutum. Dose, 5-20 min.; 0.3-1.2 cc.. —19.32; Water, to 100. Contains 17.44% by weight of real acid.

Incompatibles, readily oxidisable substances, alkalies, carbonates, iodides, bromides, chlorates, sulphides.

[2]Acidum Nitro-hydrochloricum Dilutum (Aqua Regia). Dose, 5-20 min.; 0.3-1.2 c.c. —9.38; Hydrochloric acid, 12.5; Water, 78.12. A solution containing chlorine, hydrochloric, nitric and nitrous acids.

Incompatibles, alkalies, carbonates, iodides, bromides, chlorates, sulphides, lead and silver salts.

*Acidum Oleicum. Oleic Acid.*—A straw-coloured liquid, with a faint smell, and a weak acid reaction. Insoluble in water, soluble in alcohol, ether, chloroform, fats and oils.

ACIDUM PHOSPHORICUM CONCENTRATUM. A colourless, syrupy liquid with an acid taste and reaction. Contains 66.3% of hydrogen orthophosphate, $H_3PO_4$. Sp. Gr. 1.5.

[1]Acidum Phosphoricum Dilutum.—Dose, 5-20 min.; 0.3-1.2 c.c. —15; Water, to 100. Contains 13.8% of real acid.

Incompatibles, alkalies, carbonates, lead, silver and calcium salts.

Acidum Salicylicum (see Salicin).

**Acidum Sulphuricum.** A colourless, corrosive, intensely acid liquid containing 98% by weight of hydrogen sulphate $H_2SO_4$. Sp. Gr. 1.843.

[1]Acidum Sulphuricum Dilutum.—Dose, 5-20 min.; 0.3-1.2 c.c. —8.27; Water, to 100.  Contains 13.65% of real acid.

Incompatiblies, salts of lead, barium, calcium (sulphates precipitated); bromides, iodides, chlorates.

[1]Acidum Sulphuricum Aromaticum.—Dose, 5-20 min.; 0.3-1.2 c.c. —6.98; Tincture of Ginger, 23.25; Spirit of Cinnamon, 1.18; Alcohol 68.59. The acid should be added slowly to the alcohol and the other ingredients added subsequently.

Incompatibles as for the dilute acid and also water which in large proportion precipitates the aromatics.

*Acidum Sulphurosum. Sulphurous Acid.*—Dose, 30-60 min.; 2-4 c.c. A colourless liquid with a pungent sulphurous odour, containing 6.4% of hydrogen sulphite, $H_2SO_3$.

Incompatibles, reduces chlorates, permanganates, chromates, and arsenates; silver, mercuric, and mercurous nitrates; iodides, bromides, chlorates.

**Acidum Tannicum. Tannic Acid.** (Tannin). Dose, 2-5 grs.; 1-3 dgms.
$C_{14}H_{10}O_9$. A light brownish powder consisting of thin, glistening scales, with a characteristic odour, and an astringent taste. Soluble 2 in 1 of water 10 in 6 of alcohol, and slowly 1 in 3 of glycerin.

Incompatibles, albumin, gelatin, alkaloids, alkalies, chlorates, salts of iron, lead, antimony, silver, mineral acids and lime water.

[1]Glycerinum Acidi Tannici.—20; Glycerin, to 106.

[1]Suppositoria Acidi Tannici.—Each suppository contains 3 grs. of Tannic Acid.—20; Oil of Theobroma, to 100.

[3]Trochiscus Acidi Tannici.—Each lozenge contains ½ gr. Made with the Fruit Basis.

Acidum Tartaricum.—Dose, 5-20 grs.; 3-12 dgms. $(CHOH.COOH)_2$. Colourless crystals with a strongly acid taste. Soluble 1 in 1 of water, 1 in 3 of alcohol.

Incompatibles, alkaline carbonates, salts of mercury, lead, and calcium.

**Aconiti Radix. Aconite Root.** The root of Aconitum Napellus, usually 2-4 inches long, from 1½ to 3 inches in diameter above, tapering below; dark brown in colour without; white and starchy within: odour, slight: taste, at first slight, but followed by a sensation of numbness and tingling in the mouth. The important active constituent is the alkaloid aconitine.

[1]Tinctúra Aconiti. Dose, if repeated 2-5 min.; 0.1-0.3 c.c.: for a single administration, 5-15 min.; 0.3-1 c.c.
—5; alcohol 70% 100: by percolation.

[1]Linimentum Aconiti.—66.6; Camphor, 3.33; Alcohol, to 100.

[1]Aconitina. Aconitine.—(Dose, 1/600-1/300 gr.; 1/10-1/5 mgm). An alkaloid, rarely pure. Colourless crystals with the taste of the root.

[2]Unguentum Aconitinæ.—2; Oleic Acid, 16; Lard, 82.

ADEPS. LARD.—The purified fat of the hog. A soft white fatty solid, soluble in ether, with a melting point of about 100° F. (38°-40° C.).

[1]Adeps Benzoatus. Benzoated Lard.—to 100; Benzoin, 3.

ADEPS LANÆ. WOOL FAT. A purified cholesterine-fat obtained from sheep's wool. A yellow, tenacious, unctuous substance, almost inodorous. Soluble in ether and chloroform, sparingly so in alcohol. Melting point 104°-112° F. (40°-44.4° C.).

[1]Adeps Lanæ Hydrosus. Hydrous Wool Fat (Lanolin.)—70; Water, 30.

**AETHER, ETHER.** (Sulphuric Ether). Dose, if repeated 10-30 min.; 0.6-2 c.c.: for a single administration, 46-60 min.; 2.5-4 c.c.
A colourless, very volatile, inflammable liquid, with a heavy, highly inflammable vapour, which forms an explosive mixture with air. Boiling-point lower than 105° F. (40.5° C.). Entirely miscible with alcohol, chloroform and oils. Should contain 92% of ethyl oxide $(C_2H_5)_2O$, the remainder being water and ethyl alcohol.

[1]Æther Purificatus. Purified Ether. Ether from which the water and alcohol have been removed. Used for producing anæsthesia.

[1]Spiritus Ætheris.—Dose, if repeated, 20-40 min.; 1½-2½ c.c.: for a single administration, 60-90 min.; 4-6 c.c.
—35; Alcohol, 70.

[1]Spiritus Ætheris Compositus (Hoffmann's Anodyne). Dose, if repeated, 20-40 min.; 1½—2½ c.c.: for a single administration, 60-09 min.; 4-6 c.c.
—13.75; Alcohol, 195; Sulphuric Acid, 90; Distilled Water, 3.75; Sodium Bicarbonate, sufficient to almost neutralize the acid.

*Aether Aceticus. Acetic Ether.* Dose, if repeated, 20-40 min.; 1½ to 2½ c.c.: for a single administration, 60-90 min.; 4-6 c.c.
An ethereal liquid consisting of ethyl acetate, $C_2H_5COO.CH_3$, together with small amounts of ethyl alcohol. A colourless liquid with a fragrant odour. Soluble, 1 in 10 of cold water, and in alcohol.

Aetheris Nitrosi (See Spiritus Aetheris Nitrosi, p. 96).

ALCOHOL ABSOLUTUM. ALCOHOL ABSOLUTE. (Ethyl Alcohol.) $C_2H_5OH$. A very volatile and hydroscopic liquid, containing not more than 1% of water.

SPIRITUS RECTIFICATUS. ALCOHOL (96%). Rectified Spirit.) A colourless, volatile liquid with an agreeable odour and a burning taste. Contains 90% by volume, 85.65% by weight, of ethyl alcohol, $C_2H_5OH$, the remiander being water.

The official dilutions are,—70% Alcohol.—77.77; Water, 24.16.
—60% Alcohol.—66.66; Water, 35.78.
—45% Alcohol.—50; Water, 52.66.
—20% Alcohol,—22.22; Water, 79.10.

(In all cases 100 volumes will be produced.)

SPIRITUS VINI GALLICI. BRANDY. A spirituous liquid distilled from wine and matured by age, containing not less than 36.5% by weight of ethyl alcohol.

[3]Mistura Spiritus Vini Gallici. Dose, 1-2 fl. oz.; 30-50 c.c. —40; Cinnamon Water, 40; Sugar, 5; Yolks of Eggs, 10, by volume.

VINUM XERICUM. Sherry. A Spanish wine containing not less than 13.5% of ethyl alcohol.

Vinum Aurantii (see Aurantium, p. 39).

**Aloe Barbadenis, Barbadoes Aloes.** Dose, 2-5 grs.; 1-3 dgms.
The juice that flows from the transversely cut leaves of Aloe vera, Aloe Chinensis and probably other species, dried into hard masses, which are yellowish, reddish-brown, chocolate-brown or black; odour disagreeable, taste nauseous and bitter. Almost entirely soluble in a mixture of water and alcohol in equal parts, about 30% insoluble in cold water.

Incompatibles, see Aloin.

[1]Extractum Aloes Barbadensis. Dose, 1-4 grs.; ½-2½ dgms. A dried aqueous extract; 3 of the extract is equivalent to about 4 of the drug.

[3]Decoctum Aloes Compositum. Dose ½-2 fl. oz.; 15-60 c.c. —1; Myrrh, Saffron, Potassium Carbonate, of each 0.5; Extract of Liquorice, 4; Compound Tincture of Cardamons, 30; Water, to 100.

[3]Tinctura Aloes. Dose, if repeated, 30-60 min.; 2-4 c.c.: for a single administration, 1½-2 fl. dr.; 6-8 c.c.
—2.5; Liquid Extract of Liquorice, 15; Alcohol 45%, to 100.

[1]Pilula Aloes Barbadensis. Dose, 4-8 grs.; 2½-5 dgms. (in 1 or 2 pills).
—48; Hard Soap, 24; Oil of Caraway, 3; Confection of Roses, q.s. to 100 (about 24). In each pill 2 gr. of the Extract.

[1]Pilula Aloes et Ferri.—Dose, 4-8 grs.; 2½-5 dgms. (in 1 or 2 pills)·
—22; Exsiccated Ferrous Sulphate, 11; Compound Powder of Cinnamon, 33; Syrup of Glucose, q.s. to 100 (about 33). Each pill contains ½ gr. of iron salt, 1 gr. of aloes.

**Aloe Socotrina. Socotrine Aloes.** Dose, 2-5 grs.; 1-3 dgms. The juice from the transversely cut leaves of Aloe Perryi and probably other species; dried to form either viscid or hard, dark-brown masses: odour strong, not disagreeable; taste nauseous and bitter. Almost entirely soluble in 60% alcohol: about one half is soluble in water.

Incompatibles, see Aloin.

[1]Pilula Aloes Socotrinæ. Dose, 4-8 grs.; 2½-5 dgms. (in 1 or 2 pills).
—48; Hard Soap, 24; Oil of Nutmeg, 3; Confection of Roses, q.s. to 100 (about 24). Each pill contains 2 grs. of aloes.

[2]Pilula Aloes et Asafetidæ. Dose, 4-8 grs.; 2½-5 dgms. (in 1 or 2 pills).
Aloes, Asafetida, Hard Soap, Confection of Roses, of each 25.

Each pill contains 1 gr. of both the aloes and the asafetida.

[2]Pilula Aloes et Myrrhæ. Dose, 4-8 grs.; 2½-5 dgms. (in 1 or 2 pills).
—44; Myrrh, 22; Syrup of Glucose, to 100 (about 34).
Each pill contains about 2 grs. of aloes.

**Aloinum. Aloin.** Dose, ½-2 grs.; 3-12 cgms.
A crystalline active principle contained in Aloes, yellow, odourless; taste, nauseous and bitter. Slightly soluble in water, more so in alcohol, soluble in glycerin.

Incompatibles, alkaline hydrates (which decompose it); nitric acid, ferric chloride, spirits of nitrous ether (colour change occurs).

ALUMEN. ALUM.—Dose, 5-10 grs.; 3-6 dgms.
Aluminium and potassium sulphate, $Al_2(SO_4)_3$, $K_2SO_4$, 24 $H_2O$, or aluminium and ammonium sulphate, $Al_2(SO_4)_3$, $(NH_4)_2 SO_4$, 24 $H_2O$. Colourless, transparent crystals having a sweetish, astringent taste.

Soluble 1 in 10 of cold, 1 in 1/3 of hot water; soluble in glycerin; insoluble in alcohol.

Incompatibles, alkali hydrates, or carbonates, borax and lime water (citrates, tartrates, glycerin, sugar, acacia, in part prevent precipitation); phosphates, tannic acid, tartaric acid; lead, barium, mercury and iron salts.

[2]Alumen Exsiccatum.—Potassium alum from which the water of crystallization has been driven off by heat.

[2]Glycerinum Aluminis.—16.66 Water, 6.25; Glycerin, to 100.

[2]Kaolinum (See p. 69.)

*Ammoniacum. Ammoniacum.*—Dose, 5-15 grs.; 3-10 dgms.
A gum-resin exuded from the flowering and fruiting stem of Dorema Ammoniacum. It forms small pale yellowish or brownish tears or masses, hard and brittle when cold, softening when warmed. Odour, faint; taste acrid and bitter. Triturated with water forms a white emulsion; about 60% is soluble in alcohol.

Emplastrum Ammoniaci cum Hydrargyro.—164; Mercury, 41; Olive Oil, 1.75; Sublimed Sulphur, 0.25.

Mistura Ammoniaci.—Dose, ½-1 fl. oz.; 15-30 c.c.

—3; Syrup of Balsam of Tolu, 6; Water, to 100.

## AMMONIUM. AMMONIUM. NH$_3$.

Liquor Ammoniæ Fortis. Strong Solution of Ammonia.—An aqueous solution containing 33% of ammonium, NH$_3$. A colourless liquid with a pungent, suffocating odour.

[1]Liquor Ammoniæ.—Contains 10% of Ammonium.—33.33; Water, 66.66.

[2]Linimentum Ammoniæ,—25; Almond Oil, 25; Olive Oil, 50.

[1]Spiritus Ammoniæ Aromaticus. Dose 20-40 min.; 1½-2½ c.c. repeated; for a single administration, 60-90 min., 4-6 c.c.

—4; Ammonium Carbonate, 2; Oil of Nutmeg, 0.28; Oil of Lemon, 0.40; Alcohol, 60; Water, 30. A transparent nearly colourless liquid with a pungent taste and odour. Contains the equivalent of about 2.4% of ammonium (partly as carbonate).

Incompatibles, solutions of salts of lead, silver, mercury, bismuth, antimony, copper, iron, aluminium, and zinc (precipitation prevented or hindered by the presence of sugar, acacia, glycerine, citrates and tartrates); alkaloids; chloral, and thymol.

[3]Spiritus Ammoniæ Fetidus. Fetid Spirit of Ammonia. Dose, if repeated, 20-40 min.; 1½-2½ c.c.; for a single administration, 60-90 min.; 4-6 c.c.

—10; Asafetida, 7.5; Alcohol, to 100.

[3]Ammonii Benzoas.—Dose, 5-15 grs.; 3-10 dgms.
Colourless crystals. Soluble, 1 in 6 of water; 1 in 30 of alcohol; 1 in 8 of glycerin.

[1]Ammonii Bromidum.—Dose, 5-30 grs.; 3-20 dgms.
Small colourless crystals, with a somewhat pungent taste. Soluble 2 in 3 of water; 1 in 15 of alcohol.

[1]Ammonii Carbonas.—Dose, 3-10 grs.; 2-6 dgms.
A mixture of ammonium hydrogen carbonate $NH_3HCO_3$, and ammonium carbamate, $NH_4 NH_2 CO_2$. Hard translucent crystalline masses, with an ammoniacal odour and an alkaline reaction. Soluble, 1 in 4 of water.

Incompatibles, alkali hydrates and carbonates, salts of all metals in solution, many alkaloids, and resorcin.

[1]Ammonii Chloridum.—Dose, 5-20 grs.; 3-12 dgms.
Colourless, inodorous crystals, with a pungent saline taste. Soluble 1-3 of water 1 in 60 of alcohol.

Incompatibles, lead, silver, mercurous salts; alkaline hydrates.

[3]Ammonii Phosphas.—Dose, 5-20 grs.; 3-12 dgms.
Di-ammonium hydrogen phosphate, $(NH_4)_2 HPO_4$. Transparent crystals. Soluble 1 in 4 of water, insoluble in alcohol.

Incompatibles, salts of metals; alkali hydrates.

[1]Liquor Ammonii Acetatis. (Mindererus' Spirit).—Dose, 2-6 fl. dr.; 8-23 c.c.
Ammonium Carbonate, 5; Acetic Acid sufficient to neutralise the carbonate; Water to 100. A clear colourless liquid with an acetous odour and a sharp, saline taste.

[1]Liquor Ammonii Citratis.—Dose, 2-6 fl. dr.; 8-23 c.c.
Citric Acid, 12.5; Ammonium Carbonate, sufficient to neutralise the acid in solution, about 8.75; Water, to 100. A colourless, odourless liquid with a saline taste.

*Amygdala Amara. Bitter Almond.* The ripe seed of Prunus Amygdalus var. amara. Resembles the sweet almond but is shorter and bitter.

*Amygdala Dulcis. Sweet Almond.* The ripe seed of Prunus Amygdalus var. dulcis.

Pulvis Amygdalæ Compositus.—62; Sugar, 31; Gum Acacia, **7.75.**

Mistura Amygdalæ. Dose, ½-2 fl. dr.; 2-8 c.c.
—12.5; Water, to 100.

Oleum Amygdalæ. A bland, almost odourless oil prepared from either the sweet or the bitter almond. Makes a whiter ointment than olive oil.

**Amyl Nitris.** Amyl Nitrite. Dose, by inhalation, 2-5 min.; 0.1-0.3 c.c.: (½-1 min.; 0.03-0.06 c.c.).
Yellow, fragrant liquid, slightly acid, consisting chiefly of iso-amyl nitrite. Insoluble in water; soluble in alcohol, but in time forms amyl alcohol (poisonous) and ethyl nitrite.

*Amylum.* *Starch.* Starch of wheat, maize, or rice. Soluble in hot water; insoluble in alcohol.
Incompatibles, strong alcohol, tannic acid, lead subacetate.
Glycerinum Amyli.—11; Water, 16.5; Glycerin, 71.5.

*Anethi Fructus.* *Dill Fruit.* The dried ripe fruit of Pucedanum Graveolens.
Aqua Anethi.—10; Water, 200; distill over 100.
Oleum Anethi. Dose, ½-3 min.; 0.03-0.2 c.c.
A pale yellow oil, taste sweet and aromatic, odour characteristic.

*Anisi Fructus.* *Anise Fruit.* The dried ripe fruit of Pimpinella Anisum.
Aqua Anisi.—10; Water 200; distill over 100.
[1]Oleum Anisi. Dose, ½-3 min.; 0.03-0.2 c.c.
A pale yellow oil, with an aromatic taste and a characteristic odour.
[1]Spiritus Anisi. Dose, 5-20 min.; 0.3-1.2 c.c.
—10; Alcohol, 90.

*Anthemidis Flores.* *Chamomile Flowers.* The dried flower-heads of Anthemis nobilis; odour strong and aromatic, taste bitter.
Extractum Anthemidis. Dose, 2-8 grains.; 1-5 dgms.
—10; Oil of Chamomile, 0.02; Water, 100; evaporated to a soft extract.
Oleum Anthemidis. Dose ½-3 min.; 0.03-0.2 c.c.
A pale bluish oil (becoming yellow on keeping), with an aromatic odour.

**Antimonium. Antimony.**
[3]Antimonium Nigrum Purificatiom. Antimonious Sulphide. A purified native sulphide, $Sb_2 S_3$; a greyish black powder. From it are made the other preparations.
[3]Antimonii Oxidum. Antimonious Oxide. Dose, 1-2 grs.; 6-12 cgms. $Sb_4 O_6$, a greyish white powder; insoluble in water.
[3]Pulvis Antimonialis (James' Powder.). Dose, 3-6 grs.; **2-4** dgms.
—33.3; Calcium Phosphate, 66.6.

[3]Antimonium Sulphuratum. Sulphurated Antimony. Dose, 1-2 grs.; 6-12 cgms.
A mixture of sulphides and oxides, forming a dull red powder. Insoluble in water.

[2]Antimonium Tartaratum. Tartarated Antimony (Tartar Emetic). Dose, as a diaphoretic, 1/24-1/8 gr.; 3-8 mgms.: as an emetic, 1-2 grs.; 6-12 cgms.

A double salt of Antimony and Potassium Tartrate, [K(SbO)$C_4H_4$ $O_2$], $H_2O$. Colourless, transparent crystals, with a sweet metallic taste. Soluble 1 in 17 of cold, 1 in 3 of hot water; insoluble in alcohol but moderately so in weak alcohol. Soluble in a solution of the alkaline chlorides.

Incompatibles, hydrochloric, nitric, and sulphuric acids; alkali hydrates and carbonates (prevented by citrates, tartrates, glycerin, sugar and acacia); lime water; salts of most metals; tannic acid, albumin, soap.

[1]Vinum Antimoniale. Antimonial Wine. Dose, as a diaphoretic, 10-30 min.; ½-2 c.c.: as an emetic 2-4 fl. dr.; 8-15 c.c.
—0.457; boiling Water, 5.028; Sherry, to 100. 2 gr. in 1 fl. oz.

**Apomorphinæ Hydrochloridum.** Dose, 1/10-1/4 gr.; 6-16 mgms. hypodermically 1/20-1/10 gr.; 3-6 mgms.
An alkaloid derived synthetically from morphine. Small whitish crystals, turning green on exposure to light and air. Soluble, 1 in 60 of water, 1 in 50 of alcohol. The solutions are decomposed on boiling.

[1]Injectio Apomorphinæ Hypodermica. Dose, 5-10 min.; 0.3-0.6 c.c.
—1; Diluted Hydrochloric Acid, 1; Water, recently boiled 100. 1 gr. in 110 min.

AQUA DESTILLATA. DISTILLED WATER. (Referred to throughout this book simply as Water).

*Aquæ.* (See Anethum, Anisum, Aurantium, Camphora, Caruum, Chloroformum, Cinnamomum, Foeniculum, Laurocerasus, Mentha Piperita, Mentha Viridis, Pimenta, Rosa, Sambucus.)

*Araroba. Araroba.* (Goa Powder.) A brownish powder found in the trunks of Andira Araroba.

[1]Chrysarobin. Chrysarobin. A crystalline, inodorous, tasteless, yellow powder, containing varying proportions of chrysophanic acid. Soluble slightly in water, almost entirely so in hot alcohol, completely so in hot chloroform.

[2]Unguentum Chrysarobini.—4; Benzoated Lard, 96.

### Argentum. Silver.

[1]Argenti Nitras. Dose, 1/4-½ gr.; 15-30 mgms. Colourless crystals. Soluble 1 in less than 1 of water; slightly soluble in alcohol; soluble in ether and glycerin.

Incompatibles, most inorganic salts and many organic preparations.

[1]Argenti Nitras Induratus. Toughened Caustic. Opaque white cylindrical bars.—95; Potassium Nitrate, 5. Fused and poured into moulds.

[1]Argenti Nitras Mitigatus. Mitigated Caustic. Resembles the above—33; Potassium Nitrate, 66. Fused and poured into moulds.

[3]Argenti Oxidum. Dose, ½-2 grs.; 3-12 cgms. A brown powder insoluble in water and alcohol.

Incompatibles, forms explosive mixtures with sulphur, sulphides, phosphorus, tannic acid, creosote, and many other organic substances.

*Armoraciæ Radix. Horse-radish Root.* The root of Cochlearia Armoracia: nearly cylindrical, 24 inches or more in length, ½-1 inch in diameter, externally pale yellow, internally white, odour pungent when bruised or scraped, taste pungent.

Spiritus Armoraciæ Compositus. Dose, 1-2 fl. drs.; 4-8 c.c. The root macerated with water, to which Bitter Orange Peel, Nutmeg, and Alcohol are added; and distilled.

*Arnicæ Rhizoma. Arnica Rhizome.* The dried rhizome and roots of Arnica montana. The horizontal, cylindrical, dark-brown rhizome (rootstock) is 1-2 inches long, 1/6-1/4 inches in diameter, curved and rough with leaf scars above and roots or their scars below; odour faintly aromatic, taste bitter and acrid.

Tinctura Arnicæ.—20; alcohol 70%, 100: by percolation. Used externally.

### ARSENIUM, ARSENIC.

[1]Acidum Arseniosum. Arsenious Anhydride. (Arsenicums, Arsenic. White Arsenic. Arsenious Acid.) Dose, 1/60-1/15 gr.; 1-4 mgms. Arsenious anhydride, $As_4O_6$, occurs as a heavy white powder, or as stratified partially crystalline masses; tasteless, odourless, in aqueous solution slightly acid in reaction. Soluble 1 in 20 of hot water, 1 in 60 of cold; 1 in 8 of glycerin; moderately soluble in solutions of the hydrates and carbonates of the alkalies and in solutions of hydrochloric acid.

Incompatibles of arsenious anhydride and of arsenites, most metallic salts in solution, potassium iodide (1 dr. of potassium iodide in 1 dr. of

Arsenical Solution gives but a slight precipitate), mercuric chloride, tannic acid, hypophosphites in acid mixture.

[1]Liquor Arsenicalis. Arsenical Solution (Fowler's Solution). Dose, 2-8 min.; 0.1-0.5. c.c.
—1; Potassium Carbonate, 1; Compound Tincture of Lavender, 3.125; Water, to 100
A reddish liquid alkaline to litmus. 1 gr. in 110 min.

Incompatibles, as an alkaline solution with acid solutions (Liquor Strychninæ, Liquor, and Tinctura Ferri Perchloridi, etc.) alkaloids and most metals; see also above under Acidum Arseniosum.

[1]Liquor Arsenici Hydrochloricus. Hydrochloric Solution of Arsenic. Dose, 2-8 min.; 0.1-0.5 c.c.
—1; Hydrochloric Acid, 1.25; Water to 100. A colourless liquid with an acid reaction. 1 gr. in 110 min.

Incompatibles, as for arsenious and hydrochloric acids.

[3]Arsenii Iodidum. Arsenious Iodide. Dose, 1/20-1/5 gr.; 3-12 mgms.
Small orange crystals or crystalline masses. Soluble in water and in alcohol.

Incompatibles, see above and as for any soluble iodide.

[1]Liquor Arsenii et Hydrargyri Iodidi. Solution of Arsenious and Mercuric Iodides. (Donovan's Solution.) Dose, 5-20 min.; 0.3-1.2 c.c.
—1; Mercuric Iodide, 1; Water, to 100. A clear, pale yellow liquid with a metallic taste. 1 gr. of each in 110 min. Incompatible with all alkaloids.

[3]Ferri Arsenas. Dose 1/16-1/4 gr.: 4-6 mgms (See p. 59).

[3]Sodii Arsenas. Dose 1/40-1/10 gr.: 1-6 mgms. (See p. 94).

[3]Liquor Sodii Arsenatis. Dose 2-8 min.: 0.1-0.8 c.c. (see p. 94).

ASAFETIDA. ASAFETIDA. Dose, 5-15 grs.; 3-10 dgms.
A gum resin obtained from the root of Ferula fetida. Flattened tears or masses of tears, dull yellow in colour, darkening on keeping, but yellowish or milky white within; odour strong alliaceous and persistent; taste bitter, acrid and alliaceous. Forms a white emulsion when triturated with water; in part soluble in alcohol.

[3]Tinctura Asafetidæ. Dose, 30-60 min.; 2-4 c.c.
—20; Alcohol; 70% to 100: by maceration.

Pilula Aloes et Asafetidæ, see Aloe.

Atropina, see Belladonna. (p. 40).

AURANTII CORTEX RECENS. Fresh Bitter Orange Peel. The fresh outer part of Citrus Aurantium var. Bigaradia.

[1]Tinctura Aurantii. Dose, 30-60 min.; 2-4 c.c.
—25; Alcohol, to 100.

[1]Syrupus Aurantii. Dose, 30-60 min.; 2-4 c.c.
—12.5; Syrup, 87.5.

[3]Syrupus Aromaticus. Dose, 30-60 min.; 2-4 c.c. —25; Cinnamon Water, 25: Syrup, 50.

[3]Vinum Aurantii. Made by fermenting a saccharine solution to which Bitter Orange Peel has been added. Contains 10% of ethyl alcohol.

*Aurantii Cortex Siccatus. Dried Bitter Orange Peel.* The above dride
[2]Infusum Aurantii. Dose, ½-1 fl. oz.; 15-30 c.c. —5; boiling Water, to 100.

[3]Infusum Aurantii Compositum. Dose, ½-1 fl. oz.; 15-30 c.c. —2.5; Fresh Peel, 1.25; Cloves, 0.63; boiling Water, to 100.

*Aqua Aurantii Floris. Orange-flower Water.* Obtained by distilling the flowers of Citrus aurantium var. Bigaradia with water.

Syrupus Aurantii Floris. Dose, 30-60 min.; 2-4 c.c. —11.12; Sugar, 66.67; Water, to 100.

BALSAMUM PERUVIANUM. BALSAM OF PERU. Dose, 5-15 min.; 0.3-1 c.c.
A balsam exuded from the trunk of Myroxylon Pereiræ, after the bark has been beaten and scorched. A viscid liquid, in bulk black, but in thin layers deep orange-brown or reddish-brown, and transparent; odour agreeable and balsamic; taste, acrid and leaving a burning in the throat if swallowed.

Insoluble in water; soluble 1 in 1 of alcohol but made turbid by two volumes.

Important active constituents, benzyl benzoate and cinnamate.

**Balsamum Tolutanum. Balsam of Tolu.** Dose 5-15grs.; 3-10 dgms. A balsam exuded from the trunk of Myroxylon toluifera, when incised. At first a soft and tenaceous solid, it becomes hard and brittle when dried: in thin films it is transparent and yellowish-brown in colour: odour fragrant; taste aromatic and slightly acid. Important active constituents, free cinnamic acid, benzyl cinnamate and benzoate. Insoluble in water soluble in alcohol.

[1]Syrupus Tolutanus. Dose 30-60 min.; 2-4 c.c. Contains 2.62% of balsam and 66.5% of sugar by weight.

[1]Tinctura Tolutana. Dose, 30-60 min.; 2-4 c.c. —10; Alcohol 100: by maceration.

**BELLADONNÆ FOLIA.** The fresh leaves and branches of Atropa Belladonna. The leaves are 3-8 inches long; broadly ovate, acute, entire, nearly glabrous. Important constituent, the alkaloid atropine.

[1]Extractum Belladonnæ Viride. Green Extract of Belladonna. Dose, 1/4-1 gr.; 15-60 mgms.
The juice expressed from the fresh leaves, filtered and concentrated at a low heat. (A frequent ingredient of pills.)

[3]Succus Belladonnæ. Juice of Belladonna. Dose, 5-16 min.; 0.3-1 c.c.
To three parts of the fresh juice one part of alcohol is added.

**BELLADONNÆ RADIX.** The root of Ʋtropa belladonna. Cylindrical pieces, entire or longitudinally split, ½-1 foot in length, 3/8-3/4 inch in diameter, externally longitudinally wrinkled, and greyish-brown in colour, internally whitish and starchy. Important active constituent, the alkaloid, atropine.

[1]Extractum Belladonnæ Liquidum. An alcoholic extract standardized to contain 0.75% of alkaloids. 3/4 gr. in 110 min.

[1]Extractum Belladonnæ Alcoholicum. Dose, 1/4-1 gr.; 15-60 mgms.
The above evaporated and milk sugar added to form a slightly coherent powder. Standardized to contain 1% of alkaloids.

[2]Suppositoria Belladonnæ. 1/60 gr. of alkaloids in each.—10; Oil of Theobroma to 100.

[2]Emplastrum Belladonnæ. Base, Resin Plaster. Contains 0.5% of alkaloids.

[2]Linimentum Belladonnæ—50; Camphor, 5; Water, 10; Alcohol, to 100.

[1]Tinctura Belladonnæ. Dose, 5-15 min.; 0.5-1 c.c.
—6.26; Alcohol, to 100. Standardized to contain about 0.05% of alkaloids.

**ATROPINA. ATROPINE.** Dose, 1/200-1/100 gr.; 0.3-0.6 mgms Colourless crystals, in solution with a bitter taste. Soluble 1 in 300 of water; readily in alcohol, chloroform, and ether.

Incompatibles of atropine and its salts, as for alkaloids; sodium and potassium hydrates and carbonates (not bicarbonates); decomposed by heating in acid, alkaline or neutral aqueous solution.

[2]Unguentum Atropinæ.—2; Oleic Acid, 8; Lard, 90.

[1]Atropinæ Sulphas. Dose, 1/200-1/100 gr.; 1/3-2/3 mgms.
Nearly colourless crystals. Soluble, 1 in 1 of water; 1 in 10 of alcohol; insoluble in ether and chloroform.

[1]Liquor Atropinæ Sulphatis.   Dose, ½-1 min.; 0.03-0.06 c.c.
—1 ;Salicylic Acid, 1.12; Water, to 100. 1 gr. in 110 min.

[1]Lamellae Atropinæ.  Gelatin disks each containing 1/5000 gr.

**Benzoinum.  Benzoin.** A balsamic resin obtained from Styrax Benzoin, and probably other species. Flat or curved tears varying in size but seldom exceeding two inches in length and half an inch in thickness, yellowish or reddish-brown in colour externally, milky white internally, brittle when cold, softens when warmed and when heated emits fumes of benzoic acid: the tears may occur in agglutinated masses.  Insoluble in water; soluble in alcohol .

Important active constituent, benzoic acid.

[1]Adeps Benzoatus.   Benzoated Lard.—3: Lard, 100.

[1]Tinctura Benzoini Composita. (Friar's Balsam.) Dose, 30-60 min.; 2-4 c.c.
—10; Storax, 7.5; Balsam of Tolu, 2.5; Socotrine Aloes, 1.83; Alcohol, to 100.

ACIDUM BENZOICUM.   Dose, 5-15 grs.; 3-10 dgms.
Light colourless, crystalline scales or needles; odourless, or with a slight balsamic odour. Soluble, 1 in 390 of cold, 1 in 17 of boiling water; 1 in 3 of alcohol; and in ether, chloroform, fixed and volatile oils.

Incompatibles of benzoic acid and benzoates, silver, mercury, lead, and ferric salts in solution, quinine bisulphate; strong acids (free benzoic acid from benzoates leading to a precipitate if in strong solution); benzoic acid frees carbonic acid from carbonates.

[3]Trochiscus Acidi Benzoici. ½ gr. in lozenge made with the Fruit Basis.

Ammonii Benzoas (see Ammonium, p. 34).

Sodü Benzoas (see Sodium, p. 94.)

*Benzol.* A mixture of hydrocarbons containing about 70% of benzene, and 20-30% of toluene. A colourless, volatile, inflammable liquid, with a characteristic odour.

**Bismuthum.  Bismuth.**

[1]Bismuthi Carbonas.  Bismuth Oxycarbonate.  Dose, 5-20 grs.; 3-12 dgms.
$(Bi_2O_2CO_3)_2 H_2O$. A heavy white crystalline powder, faintly acid. Insoluble in water.

[3]Trochiscus Bismuthi Compositus.—2 gr.; Precipitated Calcium Carbonate, 4 gr.; Heavy Magnesium Carbonate, 2 gr.; with the Ros- Basis.

[1]Bismuthi Subnitras. Bismuth Oxynitrate. Dose, 5-20 grs.; 3-12 dgms.
$BiONO_3$, $H_2O$. A heavy white crystalline powder, faintly acid. Insoluble in water, somewhat soluble in glycerin.

Incompatibles, carbonates and bicarbonates, iodides, hypophosphites, tannic acid.

[2]Bismuthi Oxidum. Bismuth Oxide. Dose, 5-20 grs.; 3-12 dgms.
$Bi_2O_3$. A heavy brownish-yellow powder. Insoluble in water.

[2]Bismuthi Salicylas. Bismuth Oxysalicylate. Dose, 5-20 grs.; 3-12 dgms.
A nearly white powder. Insoluble in water.
Like all salicylates gives a violet colour with ferric salts.

[2]Liquor Bismuthi et Ammonii Citratis. (Liquor Bismuthi.) Dose, 30-60 min.; 2-4 c.c.
A colourless, slightly alkaline solution, containing 3 grs. of Bismuth Oxide in 60 min.

Incompatibles, most mineral acids and the stronger organic acids precipitate bismuth citrate.

### Boron. Boron.

[1]Acidum Boricum. Boric Acid. (Boracic Acid.) Dose, 5-15 grs.; 3-10 dgms.
$H_3BO_3$. Colourless, pearly lamellar crystals, odourless, unctuous to the touch, slightly bitter and acrid in taste. Soluble, 1 in 30 of cold, 1 in 3 of boiling water, 1 in 4 of glycerin, 1 in 30 of alcohol.

Incompatibles, carbonates, mercuric chloride (a basic chloride is formed) silver nitrate, lead acetate, barium chloride, calcium chloride, (borates formed) alum, zinc sulphate, and ferric chloride.

[1]Glycerinum Acidi Borici.—6; Glycerin by weight to 100, heated to 302° F.

[2]Unguentum Acidi Borici.—10; White Paraffin Ointment, 90.

### Borax. Borax.
(Sodium Biborate or Pyroborate.) Dose, 5-20 grs.; 3-12 dgms.
$Na_2B_4O_7$, $10H_2O$. Colourless, transparent crystals, sometimes slightly effloresced, with a weakly alkaline reaction in solution. Soluble 1 in 25 of cold, 2 in 1 of boiling water; 1 in 1 of glycerin; insoluble in alcohol. Incompatibles, as an alkali, alkaloids and chloral; acacia and the metals mentioned above under boric acid.

[1]Glycerinum Boracis.—14.25; Glycerin, 85.5. (Contains some free acid.)

[2]Mel Boracis.—10.5; Glycerin, 5.25; Clarified Honey, 84.

BUCHU FOLIA. BUCHU LEAVES. The dried leaves of Barosma Betulina. Rhomboid ovate, yellowish-green leaves, with a denticulate margin, almost glabrous surface, with many oil glands upon it; odour and taste strong and characteristic, especially when crushed.

[1]Infusum Buchu. 1-2 fl. oz.; 30-60 c.c.
—5; boiling Water, 100.

[2]Tinctura Buchu. Dose, 30-60 min.; 2-4 c.c. (Contains alcohol 60%.)

*Butyl-chloral Hydras. Butyl-chloral.* Dose, 5-20 grs.; 3-12 dgms. Trichlorbutylidine-glycol, $CH_3CHCl.CCl_2CH(OH)_2$. Pearly white laminar crystals, with a pungent odour and an acrid nauseous taste. Soluble 1 in 50 of water; 1 in 1 of glycerine; 1 in 1 of alcohol.

Incompatibles, water (decomposes if kept in aqueous solution for a long time), alkalies (freeing chloroform); if triturated with acetamide, carbolic acid, menthol and urethane it liquifies.

**CAFFEINA. CAFFEÏNE.** (Theine). Dose, 1-5 grs.; ½-3 dgms. An alkaloid obtained from the leaves of Camellia Thea (the tea plant) and from beans of Coffea Arabica (coffee). Colourless inodorous acicular silky crystals. Soluble, 1 in 80 of cold, readily in boiling water; 1 in 40 of alcohol.

Incompatibles, tannic acid, mercuric salts. (Compatible with other alkaloidal precipitants.)

[1]Caffeinæ Citras. Dose, 2-10 grs.; 1-6 dgms.
A white odourless powder, with an acid faintly bitter taste, and an acid reaction when in solution. Solubility, 1 in 32 of water, with 3 parts of water gives a syrupy solution, 1 in 22 of alcohol.

[1]Caffeinæ Citras Effervescens. Dose, 60-120 grs.; 4-8 gms.
—4; Sodium Bicarbonate, 51; Citric acid, 18; Tartaric Acid, 27; Sugar, 14.

**Calcium. Calcium.**

Incompatibles, the hydrate, carbonate, sulphate, phosphate, oxalate, and tartrate are insoluble and in consequence soluble calcium salts are incompatible with the acids of these salts or with soluble salts of the same.

[1]Calcii Carbonas Praecipitatus. Precipitated Calcium Carbonate. (Precipitated Chalk.) Dose, 10-60 grs.; ½-4 gms.
A white microcrystalline powder. Insoluble in water.

[1]Calcii Chloridum. Dose, 5-15 grs.; 3-10 dgms.
$CaCl_2, 2H_2O$. White very deliquescent masses, with a bitter acrid taste. Soluble 1 in 1 of water; 1 in 3 of alcohol.

[2]Calcii Hydras. Calcium Hydrate. (Slaked Lime.) Ca(OH)₂. Must be freshly prepared by the action of calcium oxide and water. Soluble 1 in 900 of water.

[1]Liquor Calcis (Lime Water). Dose, 1-4 fl. oz.; 30-120 c.c. Made by shaking calcium hydroxide with water and decanting. ½ gr. of lime in 1 fl. oz.

[1]Linimentum Calcis. (Caron Oil.) Equal parts of Lime Water and Olive Oil shaken together.

[2]Liquor Calcis Saccharatus. Saccharated Solution of Lime. Dose, 20-60 min.; 1-4 c.c.
—5; Sugar, 10; Water to 100. Shake and decant the clear fluid. 8 grs. in 1 fl. oz.

[2]Calcii Hypophosphis. Dose, 3-10 grs.; 2-6 dgms. Ca(PH₂O₂)₂. A white powder or lustrous crystals, odourless and having a nauseous taste. Soluble 1 in 8 of water; insoluble in alcohol.

Incompatibles, as for calcium, and also chlorates, mercuric salts, bismuth subnitrate, ferric salts, quinine, strong acids, (the acid solution of arsenic); explodes if triturated with oxidizing agents.

[3]Calcii Phosphas. Dose, 5-15 grs.; 3-10 dgms. A light white amorphous powder. Insoluble in water; soluble in acetic, hydrochloric or nitric acids.

[2]Calx. Lime. (Burnt Lime. Calcium Oxide.) CaO. Compact whitish masses.

[2]Calx Chlorinata. Chlorinated Lime. A mixture containing the hydrate, chloride, and hypochlorite of calcium. A dull whitish powder with a characteristic smell. Decomposes on exposure to the air. Partially soluble in water.

[3]Liquor Calcis Chlorinatæ. (Bleaching Liquid.)—10; Water to 100.

[3]Calx Sulphurata. Sulphurated Lime· Dose, 1/4-1 gr.; 15-60 mgms.
A mixture containing about 50 % of calcium sulphide with calcium sulphate and carbon. A greyish white powder with a smell of hydrogen sulphide.

[1]Creta Preparata. Dose, 10-60 grs.; ½-4 gms. Native calcium carbonate, freed from most of its impurities by elutriation. White friable masses or white powder.

[1]Mistura Cretæ. Chalk Mixture. Dose ½-1 fl. oz.; 15-30 c.c.
—3.125; Tragacanth, 0.44.; Sugar, 6.25; Cinnamon Water to 100. In 1 fl. oz. 14 grs. of chalk.

[1]Pulvis Cretæ Aromaticus. Dose, 10-60 grs.; ½-4 gms.
—24.2; Cinnamon Bark, 8.8; Nutmeg, 6.6; Cloves, 3.3; Cardamons, 2.2; Sugar, 55.

[1]Pulvis Cretæ Aromaticus cum Opio. Dose, 10-40 grs.; 6-25 dgms.
—97.5; Opium, 2.5.  1 gr. of Opium in 40 grs.

[3]Syrupus Calcii Lactophosphatis. Dose, 30-60 min.; 2-4 c.c. Precipitated Calcium Carbonate, 2.5; Lactic Acid, 6.0; Phosphoric Acid Concentrated, 4.6; Sugar, 70.0; Orange-flower Water, 2.5; Water to 100.

**Calumbæ Radix. Calumba Root.** The dried transversely cut slices of the root of Jateorhiza Calumba. In irregular flattish, roughly circular, yellowish slices, about 1-2 or more inches in diameter, and 1/8-½ or more of an inch thick, odour feeble, taste bitter. As this bitter contains no tannin it may be given with iron, acids, or alkalies.

[1]Infusum Calumbæ. Dose, ½-1 fl. oz.; 15-30 c.c.
—5; Water, 20.

[3]Liquor Calumbæ Concentratus. Dose, 30-60 min.; 2-4 c.c.
—50; Alcohol, 22.5; Water to 100: by maceration.

[1]Tinctura Calumbæ. Dose, 30-60 min.; 2-4 c.c.
—10; Alcohol 60% to 100: by maceration.

*Cambogia. Gamboge.* Dose, ½-3 grs.; 3-12 cgms.
A gum-resin obtained from Garcinia Hamburii. Cylindrical, longitudinally striated rolls, often agglutinated into masses; colour reddish-yellow; odourless; taste acrid. Insoluble in water but forms with it an emulsion; soluble in weak alcohol.

Pilula Cambogiæ Composita. Dose, 4-8 grs.; 2½-5 gms. (in 1 or 2 pills).
—16.5; Barbadoes Aloes, 16.5; Compound Cinnamon Powder, 16.5; Hard Soap, 33.0; Syrup of Glucose, q.s. (about 16). Each pill contains about 2/3 gr. of aloes and of gamboge.

**Camphora. Camphor.** Dose, 2-5 grs.; 1-3 dgms.
A white crystalline substance obtained from Cinnamomum Camphora, purified by sublimation. Solid, colourless, transparent, crystalline, pieces of tough consistence; with a powerful penetrating odour, and a bitter pungent taste, followed by a sensation of coldness, inflammable, burning with a smoky flame, and volatile. Soluble 1 in 700 of water, 1 in 1 of alcohol, 1 in 4 of olive oil.

Incompatibles, forms a liquid, if triturated, with phenol, chloral, menthol, thymol, salol.

[1]Aqua Camphoræ. (Dose, 30-60 c.c.)—0.1 dissolved in a little alcohol and slowly added to 100 of water.

[1]Linimentum Camphoræ. (Camphorated Oil.)—20; Olive Oil, 80.

¹Linimentum Camphoræ Ammoniatum.—12.5; Oil of Lavender, 0.625; Strong Solution of Ammonia, 25; Alcohol to 100.

¹Spiritus Camphoræ. Dose, 5-20 min.; 0.3-1.2 c.c. —10; Alcohol to 100.

¹Tinctura Camphoræ Composita. (Paregoric Elixer, Paregoric.) Dose, 30-60 min.; 2-4 c.c. —0.34; Tincture of Opium, 6.09; Benzoic Acid, 0.46; Oil of Anise, 0.31; Alcohol to 100. 1 fl. dr. contains 1/4 gr. of Opium or 1/30 gr. of morphine.

CANNABIS INDICA. INDIAN HEMP. The dried flowering or fruiting tops of the female plant of Cannabis sativa. Usually compressed masses of leaves, stems, and flowers. The leaves bear numerous oil glands and curved hairs. The active constituent seems to be cannabinnol contained in the resin.

¹Extractum Cannabis Indicæ. Dose, 1/4-1 gr.; 15-60 mgms. An alcoholic extract of a soft consistence.

¹Tinctura Cannabis Indicæ. Dose, 5-15 min.; 0.3-1 c.c. —5; Alcohol to 100.

**Cantharis. Cantharides.** The dried beetle Cantharis vesicatoria. About 3/4 to 1 inch long and a 1/4 inch broad, with two long wing-sheaths of a coppery-green colour; odour strong and disagreeable. The active principle is cantharidin, an acid anhydride, which is volatile; soluble slightly in water, less so in alcohol (1 in 1150) and in ether (1 in 700), more so in chloroform (1 in 65) and in acetone (1 in 40) and in glacial acetic acid; incompatible, with alkalies (forming salts which are active) and lead, silver, copper and mercuric salts in solution.

³Acetum Cantharidis. Vinegar of Cantharides.—10; a mixture of Glacial Acetic Acid and Water in equal parts to 100.

¹Emplastrum Cantharidis.—35; Yellow Beeswax, 20; Lard, 20; Resin, 20; Soap Plaster, 5.

²Emplastrum Calefaciens. Warming Plaster.—4; Yellow Beeswax, 4; Resin, 4; Resin Plaster, 52; Soap Plaster, 32; Water, boiling, 20.

¹Liquor Epispasticus. Blistering Liquid.—50; Acetic Ether, 100: by percolation.

¹Collodium Vesicans. Blistering Collodium.—to 100; Pyroxylin, 2.5.

²Tinctura Cantharidis. Dose, if repeated. 2-5 min.; 0.1-0.3 c.c.: for a single administration, 5-15 min.; 0.3-1 c.c. —1.25. Alcohol, 100: by maceration.

¹Unguentum Cantharidis.—10; Benzoated Lard 90.

*Caoutchouc. India-Rubber.* (Para Rubber.) The prepared milk-juice of Hevea brasiliensis and probably other species. Brownish-black, elastic masses. Insoluble in water and alcohol; soluble in chloroform, carbon bisulphide, turpentine, benzol and petroleum.

Liquor Caoutchouc.—5; Benzol,, 50; Carbon Bisulphide, 50.

CAPSICI FRUCTUS. CAPSICUM. The dried ripe fruit of Capsicum minimum. Dull orange-red, oblong fruits, about ½-¾ of an inch in length and ¼ inch in diameter; odour characteristic, taste intensely pungent.

[1]Tinctura Capsici. Dose, 5-15 min.; 0.3-1 c.c. —5; Alcohol 70% to 100 by maceration.

[3]Unguentum Capsici.—24; Spermaceti, 12; Olive Oil by weight, 88.

*Carbo Ligni. Wood Charcoal.* Dose, 60-120 grs.; 4-8 gms. The carbonaceous residue of wood charred by exposure to red heat without access of air.

*Carbonis Bisulphidum. Carbon Bisulphide.* $CS_2$. A clear colourless liquid with a characteristic but not fetid odour. Very slightly soluble in water, soluble in alcohol, ether, and chloroform, and in fixed and volatile oils.

**Cardamomi Semina. Cardamon Seeds.** The dried ripe seed: of Elettaria Cardamomum. Usually kept in their pericarps until wanteds these are ovoid or oblong, bluntly triangular in section, longitudinally striated, pale buff in colour. The seeds are brown, about 1/8 inch in length, breadth and thickness, irregularly angular, and wrinkled. Odour and taste warm and aromatic.

[1]Tinctura Cardamomi Composita. Dose, 30-60 min.; 2-4 c.c. —1.25 Caraway Fruit, 1.25; Raisins, 10; Cinnamon Bark, 2.5 ; Cochineal Bark, 0.63; Alcohol 60% 100: by maceration.

*Carui Fructus. Caraway Fruit.* The dried fruit of Caruum Carvi. The active constituent is the oil.

Aqua Carui. (Dose, 30-60 c.c.) —10; Water, 200; distil over 100.

[1]Oleum Carui. Dose ½-3 min.; 0.03-0.2 c.c. A colourless or pale yellow oil with a characteristic odour and spicy, pungent taste.

*Caryophyllum*. *Cloves*. The dried flower buds of Eugenia caryophyllata. The principle active constituent is the oil.

Infusum Caryophylli.  Dose, ½-1 fl. oz.; 15-30 c.c.
—2.5 in 100 of water.  Contains some tannin.

[1]Oleum Caryophylli.  Dose, ½-3 min.; 0.03-0.2 c.c.
A colourless or pale yellow oil becoming reddish on standing; odour and taste of cloves. Incompatibles, ferric chloride, lime water, strong alkaline or mineral acid solutions.

**Cascara Sagrada. Cascara Sagrada.** (Rhamni Purshiani Cortex' Sacred Bark.) The dried bark of Rhamnus Purshianus. Quilled, channelled or curved pieces, frequently 4 inches long, ¾ of an inch wide, and about 1/16 inch thick, the outer surface is smooth, dark purplish-brown in colour but often covered with a whitish coat of lichens; odour characteristic; taste nauseous and persistently bitter.

[1]Extractum Cascaræ Sagradæ.  Dose, 2-8 grs.; 1-5 dgms.
A dried aqueous extract.

[1]Extractum Cascaræ Sagradæ Liquidum.  Dose, 30-60 min.; 2-4 c.c.
An aqueous extract. Incompatibles, gives a precipitate with alcohol (of unimportant constituents), acids and strong solutions of mineral salts.

[3]Syrupus Cascaræ Aromaticus.  Dose, ½-2 fl. dr.; 2-8 c,c.
—40; Tincture of Orange, 10; Alcohol, 5; Cinnamon Water, 15; Syrup, 30.

*Cascarilla*. *Cascarilla*. The dried bark of Croton Eluteria. In quills from 1-3 inches long, 1/6-½ inch in diameter, or curved pieces. The outer layer is wrinkled longitudinally with transverse cracks; dull brown or dark gray in colour but frequently covered with silver gray patches containing black spots: odour, aromatic and agreeable, especially when burned; taste, aromatic and bitter.

Infusum Cascarillæ.  Dose, ½-1 fl. oz.; 15-30 c.c.
—5 in 100 of boiling water.

Tinctura Cascarillæ.  Dose, 30-60 min.; 2-4 c.c.
—20; Alcohol 70 % to 100 :by percolation.

*Cassiæ Pulpa*. *Cassia Pulp*. The pulp obtained from the pods of Cassia Fistula. The pods are blackish brown in colour, very hard, and from 1½-2 feet long. The pulp is viscid and nearly black, with a faint odour and a sweet taste. It is a constituent of Confection of Senna.

*Catechu. Catechu.* Dose, 5-15 grs.; 3-10 dgms.
An extract obtained from the leaves and young shoots of Uncaria Gambier. In cubes, about an inch on the side, often agglutinated, deep reddish brown externally, porous and friable, consisting largely of minute crystals; taste, at first bitter and astringent, subsequently sweetish; odourless. Active principles catechin and catechu-tannic acid.

Incompatibles, gelatin, albumin, sulphuric acid and ferric salts.

[1]Pulvis Catechu Compositus. Dose, 10-40 grs.; ½-2½ gms.
—40; Kino, 20; Krameria Root, 20; Cinnamon Bark, 10; Nutmeg, 10.

[2]Tinctura Catechu. Dose, 30-60 min.; 2-4 c.c.
—20; Cinnamon Bark, 5; Alcohol 60%, 100: by maceration.

[3]Trochiscus Catechu. 1 gr. in a lozenge made with the Simple Basis.


*Cera Alba. White Beeswax.* Yellow Beeswax bleached. Hard, nearly white, translucent masses.

*Cera Flava. Yellow Beeswax.* Prepared from the comb of the bee, Apis mellifica. Insoluble in water, 3% soluble in alcohol, 50% soluble in ether, completely soluble in oil of turpentine.


CERIUM. CERIUM.

[1]Cerii Oxalas. Cerium oxalate. Dose, 2-10 grs.; 1-6 dgms.
$Ce_2 (C_2O_4)_3, 9H_2O$. A nearly white, granular powder, odourless, tasteless. Insoluble in water and other ordinary solvents, soluble in dilute sulphuric and hydrochloric acids. Incompatibles, alkali hydrates.


*Cetaceum. Spermaceti.* A concrete fatty substance obtained from the head of the Sperm Whale, Physeter macrocephalus, and subsequently purified. Crystalline, pearly white, glistening, translucent masses, unctuous to the touch, odourless, and flavourless, Insoluble in water, almost so in alcohol, soluble in ether, chloroform, and boiling alcohol.

Unguentum Cetacei.—20; White Beeswax, 8 Almond Oil by weight, 72; Benzoin, 2.


*Charta* (see Sinapis p. 93).


*Chirata. Chiretta.* The dried plant Swertia Chirata, collected when in flower. Stem 3 feet or more in length, purplish externally, with a pith within; branches, slender: leaves glabrous and entire; flowers small, numerous and in panicles: odourless; taste extremely bitter. Contains a bitter but no tannic acid.

Infusum Chiratæ. Dose, ½-1 fl. oz.; 15-30 c.c.
—5; boiling Water, 100.

Liquor Chiratæ Concentratus. Dose, 30-60 min.; 2-4 c.c. Alcoholic; by percolation.

Tinctura Chiratæ. 30-60 min.; 2-4 c.c.
—10; Alcohol 60% to 100: by percolation.

**CHLORAL HYDRAS. CHLORAL HYDRATE.** (Chloral.) Dose, 5-20 grs. 3-12 dgms.
$CCl_3.CH(OH)_2$. Colourless, nondeliquescent plates; pungent odour; pungent, bitter taste. Soluble 5 in 1 of alcohol, 4 in 1 of water; or 2 in 1 of ether, soluble 1 in 4 of chloroform.

Incompatibles, alkaline hydrates, carbonates, borates, ammonia, mercuric oxide, potassium permanganate and iodide; gives a stiff mass or a liquid when triturated with phenol, lead acetate, phenacetin, salol, sodium phosphate, thymol, trional, urethane, or quinine sulphate: gives a damp powder with acetanelid or phenazone.

[1]Syrupus Chloral. Dose, ½-2 fl. dr.; 2-8 c.c.
—18.29; Water, 18.75; Syrup to 100. 10 grs. of chloral in 1 fl. dr.

**CHLOROFORMUM. CHLOROFORM.** Dose, 1-5 min.; 0.05-0.3 c.c. Chloroform is trichlormethane, $CHCl_3$, to which about 1% of absolute alcohol has been added. A heavy, colourless liquid, with a characteristic odour and a sweetish burning taste. Soluble 1 in 200 of water, miscible in all proportions with alcohol, ether, and oils.

[1]Aqua Chloroformi.—1; Water, to 400.

[1]Linimentum Chloroformi.—50; Camphor Liniment, 50.

[1]Spiritus Chloroformi. Dose, if repeated, 5-20 min.; 0.3-1.2 c.c.: for a single administration, 30-40 min.; 2-3 c.c.
—5; Alcohol, to 100.

[1]Tinctura Chloroformi et Morphinæ Composita. Dose, 5-15 min.: 0.3-1 c.c.
—7.5; Morphine Hydrochloride, 1.0; Diluted Hydrocyanic Acid, 5.0; Tincture of Capsicum, 2.5; Tincture of Indian Hemp, 10.0; Oil of Peppermint, 0.15; Glycerin, 25.0; Alcohol to 100. In 10 min. there are ¾ min. of Chloroform, 1/11 gr. of Morphine, ½ min. of Diluted Hydrocyanic Acid.

Chrysarobinum. (See Araroba p. 36).

*Cimicifugæ Rhizoma. Cimicifuga.* The dried rhizome and roots of Cimicifuga racemosa. A hard roughly cylindrical rhizome bearing numerous remains of branches encircled by leaf-scars, and the remains of brittle roots; odour faint, taste bitter and acrid. It contains tannic acid.

Extractum Cimifugæ Liquidum.  Dose, 5-30 min.; 0.3-2 c.c.
—Alcoholic.

Tinctura Cimicifugæ.  Dose, 30-60 min.; 2-4 c.c.
—10; Alcohol 60%, 100: by percolation.

## CINCHONÆ RUBRÆ CORTEX.  RED CINCHONA BARK.
The dried bark of the stem and branches of Cinchona succiruba. Quilled more or less incurved pieces, from 2-12 inches in length and 1/10 to ¼ inch thick; the outer surface brownish, roughened by numerous ridges, warts and cracks; the inner surface striated, brick-red or reddish-brown: taste bitter and somewhat astringent, no marked odour. The most important ingredient is the alkaloid quinine, other alkaloids are cinchonidine, cinchonine, and quinidine. It also contains a tannic acid.

Incompatibles, alkalies and their carbonates, tannin, ammonia, lime water, gelatin, many metallic salts (especially ferric salts).

[1]Extractum Cinchonæ Liquidum.  Dose, 5-15 min.; 0.3-1 c.c. An acid alcoholic extract standardized to contain 5% of alkaloids.

[2]Infusum Cinchonæ Acidum.  Dose, ½-1 fl. oz.; 15-30 c.c.
—5; boiling Water, 100; Aromatic Sulphuric Acid, 1.25.

[2]Tinctura Cinchonæ.  Dose, 30-60 min.; 2-4 c.c.
—20; Alcohol, 70%, 100: by percolation. Standardised to contain 1% of alkaloids.

[1]Tinctura Cinchonæ Composita.  Dose, 30-60 min.; 2-4 c.c.
—50; Bitter Orange Peel, 5; Serpentary Rhizome, 2.5; Saffron, 0.63; Cochineal, 0.32; Alcohol 70% to 100: by maceration. Standardised to contain ½% of alkaloids.

## QUININÆ SULPHAS.  QUININE SULPHATE.  Dose, 1-10 grs.;
½-6 dgms.
Filiform, silky crystals, with an intensely bitter taste. Soluble 1 in 800 of water, 1 in 65 of alcohol. The addition of roughly 1 min. per gr. of a diluted mineral acid will convert it into the acid sulphate (or bisulphate), which is soluble 1 in 10 of water.

Incompatibles, acetates, citrates, benzoates, salicylates, tartarates, alkali hydrates or carbonates, borax, tannic acid, mercuric chloride, potassium and mercuric iodides: gives a soft mass with thymol, and a stiff mass or damp powder with chloral.

[1]Pilula Quininæ Sulphatis.  Dose, 2-8 grs.; 2½-5 dgms. (in 1 or 2 pills).
—82.5; Tartaric Acid, 2.75; Glycerin, 11; Tragacanth, 2.75.

[3]Tinctura Quininæ Ammoniata.  Dose, 30-60 min.; 2-4 c.c.
—2; Solution of Ammonia, 10: Alcohol 60%, 90.

**QUININÆ HYDROCHLORIDUM.** Dose, 1-10 grs.; ½-6 dgms. Silky filiform crystals, larger than those of the sulphate. Soluble, 1 in 35 of water, 1 in 3 of alcohol. Incompatibles, as for the sulphate.

²Tinctura Quininæ. Dose, 30-60 min.; 2-4 c.c. —2; Tincture of Orange, 100: by solution.

²Vinum Quininæ. Dose, ½-1 fl. oz.; 15-30 c.c. —0.228; Orange Wine, 100.

**QUININÆ HYDROCHLORIDUM ACIDUM.** Dose, 1-10 grs.; ½-6 dgms.
A white crystalline powder. Soluble, 1 in less than 1 of water giving a slightly acid liquid. Incompatibles, as for the sulphate.

CINNAMOMI CORTEX. CINNAMON BARK. The dried inner bark of shoots from the truncated stocks of Cinnamomum zeylanicum. Closely rolled quills about 3/8 of an inch in diameter; thin splintery, light yellow-brown externally, darker brown internally; odour fragrant and characteristic, taste warm, sweet and aromatic. Contains a volatile oil and tannic acid.

¹Aqua Cinnamomi. 10 in 100 of water by distillation.

¹Oleum Cinnamomi. Dose, ½-3 min.; 0.06-0.2 c.c.
A pale yellow oil becoming reddish on standing, odour and taste of cinnamon.

¹Spiritus Cinnamomi. Dose, 5-20 min.; 0.3-1.2 c.c. —10; Alcohol, 90.

¹Pulvis Cinnamomi Compositus. (Pulvis Aromaticus.) Dose, 10-40 grs.; ½-2½ gms.
—33.3; Cardamon Seeds, 33.3; Ginger, 33.3.

²Tinctura Cinnamomi. Dose, 30-60 min.; 2-4 c.c. —20; Alcohol 70% 100: by percolation.

**Cocæ Folia. Coca Leaves.** The dried leaves of Erythroxylum Coca. Brownish-green leaves, 1½-3 inches long, entire, oval and glabrous; the mid-rib projects as a rib from the dorsal surface, and ends in a spine; on the under surface a curved line may usually be seen on either side of the mid-rib. The leaves have a faint odour and a bitter taste succeeded by a sensation of numbness. The important active principle is the alkaloid cocaine.

²Extractum Cocæ Liquidum. Dose, 30-60 min.; 2-4 c.c.
An alcoholic extract.

**COCAINA. COCAINE.** (Dose, 1/20-½ gr.; 3-30 mgms.)
Colourless prismatic crystals, with a bitter taste followed by a sensa-

tion of numbness. Almost insoluble in water, soluble 1 in 10 alcohol. 1 in 12 of olive oil, insoluble in glycerin.

[3]Unguentum Cocainæ.—4; Oleic Acid (by weight) 16; Lard, 80.

**COCAINÆ HYDROCHLORIDUM.** Dose, 1/5-½gr.; 12-30 mgms Colourless crystals, taste bitter, followed by a sensation of numbness. Soluble, 2 in 1 of water, 1 in 4 of alcohol, 1 in 4 of glycerin, insoluble in olive oil or ether. Incompatibles, as for alkaloids; and strong solutions of acids, or alkalies; calomel.

[1]Injectio Cocainæ Hypodermica. Dose, 2-5 min.; 0.12-0.3 c.c. —10; Salicylic Acid, 1.5; Water, recently boiled, 100. 1. gr. in 11 min.

[1]Lamellæ Cocainæ. Gelatine. disks each containing 1/50 gr. of cocaine.

Trochiscus Krameriæ et Cocainæ.—1/20 gr.; Krameria, 1 gr.; with the Fruit Basis.

*Coccus. Cochineal.* The dried fecundated female insect, Coccus Cacti. About 1/5 of an inch long; roughly oval in outline, transversely wrinkled, concave beneath, convex above, purplish-gray in colour; when powdered, dark red. Contains a colouring principle, carmine.

Tinctura Cocci. Dose, 5-15 min.; 0.3-1 c.c. —10; Alcohol 45%, 100: by maceration.

Codeina (see Opium p. 80).

COLCHICI CORMUS. COLCHICUM CORM. Dose, 2-5 grs.; 1-3 dgms. The fresh corm is stripped of its coats, sliced transversely and dried. The dried slices are about 1/10 of an inch thick and about 1 broad, somewhat reniform in shape, whitish in colour; taste bitter; without odour; contains an alkaloid, colchicine, which is incompatible with iodides, guaiacum, and all astringent preparations.

[3]Extractum Colchici. Dose, 1/4-1 gr.; 15-60 mgms. The juice of the fresh corms, expressed and dried to a soft consistence.

[1]Vinum Colchici. Dose, 10-30 min.; ½-2 c.c. Dried Corm, 20; Sherry, 100: by maceration.

COLCHICI SEMINA. COLCHICUM SEEDS. The dried ripe seeds of Colchicum autumnale. About 1/10 inch in diameter, reddish-brown, rough, minutely pitted; very hard and tough; odourless, taste acrid and bitter. Contains the alkaloid, colchicine. Incompatibles, as above.

[2]Tinctura Colchici Seminum. Dose, 5-15 min.; 0.3-1 c.c. —20; Alcohol 45%, 100: by percolation.

Collodium (see Pyroxylin p. 87).

**Colocynthidis Pulpa.  Colocynth Pulp.** The dried pulp of the fruit of Citrullus Colocynthis, freed from its seeds. The pulp is light, spongy, whitish, odourless, intensely bitter.

[1]Extractum Colocynthidis Compositum.  Dose, 2-8 grs.; 1-5 dgms.
A tincture of colcynth is made; the alcohol is evaporated off, the Extract of Barbadoes Aloes, Scammony Resin, Curd Soap and Cardamom Seeds are added and the whole evaporated to a firm extract.

[1]Pilula Colocynthidis Composita.  Dose, 4-8 grs.; 2½-5 dgms. (in 1 or 2 pills)—18; Barbadoes Aloes, 36; Scammony Resin, 36; Potassium Sulphate, 4.5; Oil of Cloves, 4.5; Water, q.s.  Each pill contains 3/4 gr. of Colocynth, and 1½ grs. of both Scammony and Aloes.

[1]Pilula Colocynthidis et Hyoscyami.  Dose, 4-8 grs.; 2½-5 dgms (in 1 or 2 pills).
—66; Green Extract of Hyoscyamus, 33.

*Confectiones.*  (See Piper, Rosa, Senna, Sulphur).  Dose, 60-120 grs.; 4-8 gms.

*Conii Folia. Conium Leaves.* The fresh leaves and young branches of Conium maculatum, collected when the fruit begins to form. Stem smooth marked with dark purple spots, leaves large, pinnately divided, the lower decompound and at times 2 feet long; odour strong and mouse-like, especially if rubbed with potassium hydrate. The chief active principle is the alkaloid, coniine.

Succus Conii.  Dose, 1-2 fl. dr.; 4-8 c.c.
—66 of Juice expressed from the fresh leaves and branches with 33 of alcohol.

Unguentum Conii.—200 evaporated to 25; Hydrous Wool Fat, 75.

*Conii Fructus. Conium Fruit.* (Hemlock Fruit.) The dried full-grown unripe fruit of Conium maculatum. Broadly ovoid, greenish-gray, about 1/8 inch long and broad; odour not marked unless rubbed with potassium hydrate, when strong and mouse-like. Active principle the alkaloid coniine.

Tinctura Conii.  Dose, 30-60 min.; 2-4 c.c.
—20; Alcohol 70%, 100 by percolation.

**Copaiba.  Copaiba.**  Dose, 30-60 min.; 2-4 c.c.
The oleo-resin obtained from the trunk of Copaifera Lansdorfii and probably other species. A more or less viscid liquid, generally transparent, light yellow to pale yellow-brown in colour; odour aromatic and peculiar;

taste persistent, acrid and somewhat bitter. Soluble in absolute alcohol.

Incompatibles, hydrates of the alkalies and alkaline earths.

[1]Oleum Copaibæ. Dose, 5-20 min.; 0.3-1.2 c.c.
A colourless or pale yellow oil, with the odour and taste of copaiba. Soluble 1 in 1 of absolute alcohol.

*Coriandri Fructus. Coriander Fruit.* Dose 30-60 gr: 2-4 gms. The dried ripe fruit of Coriandrum sativum. Nearly globular, about 1/5 inch in diameter, brownish-yellow in colour and glabrous; odour aromatic; taste agreeable.

Oleum Coriandri. Dose ½-3 min.; 0.03-0.2 c.c.
A pale yellow or colourless oil, with the taste and odour of the fruit.

Creta (See Calcium, p. 44).

**Creosotum, Creosote.** Dose, 1-5 min.; 0.05-0.3 c.c.
A mixture of guaiacol, cresol and other phenols. A colourless or yellowish liquid, with an empyreumatic odour and acrid taste. Soluble 1 in 150 of cold water, more soluble in hot, readily soluble in alcohol, ether, chloroform and glycerin.

Incompatibles, many metal salts, such as those of silver and copper, albumin, ferric salts, nitric acid; explodes if triturated with oxidising agents.

[3]Mistura Creosoti. Dose, ½-1 fl. oz.; 15-30 c.c.
—0.2; Spirit of Juniper, 0.2; Syrup, 6; Water to 100.

[3]Unguentum Creosoti.—10; Hard Paraffin 40; Soft White Paraffin 50.

*Crocus. Saffron.* The dried stigmas and tops of the styles of Crocus sativus. The flower parts have an aromatic odour and a bitter taste; they leave, if moistened and rubbed on the finger, an intense yellow stain.

Tinctura Croci. Dose, 5-15 min.; 0.3-1 c.c.
—5; Alcohol 60%, 100 by maceration.

Croton (see Oleum p. 76).

CUBEBÆ FRŪCTUS. CUBEBS. Dose, 30-60 grs.; 2-4 gms.
The dried full-grown unripe fruits of Piper Cubeba. Nearly globular, about 1/6 inch in diameter, greyish-brown or nearly black in colour; odour strong, aromatic and characteristic; taste warm, somewhat bitter and aromatic.

[1]Oleum Cubebæ. Dose, 5-20 min.; 0.3-1.2 c.c.
A colourless or pale greenish oil, with the odour and taste of cubebs.

[2]Tinctura Cubebæ. Dose, 30-60 min.; 2-4 c.c.
—20; Alcohol 100: by percolation.

Cuprum. Copper.

[1]Cupri Sulphas. Copper Sulphate. Dose, as an astringent, 1/4-2 grs.; 15-120 mgms.: as an emetic, 5-10 grs.; 3-6 dgms. $CuSO_4, 5H_2O$. Blue crystals. Soluble 1 in 3.5 of water (giving an acid solution), very soluble in glycerin, insoluble in alcohol.

Incompatibles, alkaline hydrates and carbonates, ammonia, phosphates, arsenites, iodides, tannic acid, albumins; in the presence of alkalies arsenious acid, glucose and acacia.

*Cuspariæ Cortex. Cusparia Bark.* The dried bark of Cusparia febrifuga. Flattened or curved pieces or quills, 4 or 5 inches long, an inch wide and a twelfth thick; the outer layer grey or yellowish, easily removed exposing the inner layer which is hard and dark brown: odour musty; taste bitter.

Infusum Cuspariæ. Dose, 1-2 fl. oz.; 30-60 c.c. —5; boiling Water, 100.

Liquor Cuspariæ Concentratus. Dose, 30-60 min.; 2-4 c.c. —50; Alcohol 20% to 100; by percolation.

Cusso. Kousso. Dose, ¼-½ oz.; 7-14 gms. The dried panicles of pistolate flowers of Brayera anthelmintica. Usually in more or less cylindrical rolls, 1-2 feet long, composed of reddish panicles of numerous small flowers: odour not marked; taste bitter and acrid.

Decocta (see Aloe, Grenatum, Haematoxylum). Dose, ½-2 fl. oz.; 15-30 c.c.

**DIGITALIS FOLIA. DIGITALIS LEAVES.** Dose, ½-2 grs.; 3-12 cgms.
The dried leaves of Digitalis purpurea. From 4-12 inches or more in length, and at times 5-6 inches broad; upper surface, rugose, dull green and slightly hairy, under surface paler and densely pubescent; no marked odour; taste very bitter. The chief active principles are the glucosides, digitalin, digitoxin, and digitalein.

Incompatibles, strong alkalies, acids, (e.g. in Tinct. Ferri Perchloridi) lead acetate, ammonia.

[1]Infusum Digitalis. Dose, 2-4 fl. dr.; 8-16 c.c. —0.68; boiling Water, 100.

[1]Tinctura Digitalis. Dose, 5-15 min.; 0.3-1 c.c. —12.5; Alcohol 60%, 100: by percolation.

**Elaterium. Elaterium.** Dose, 1/10-½ gr.; 6-30 mgms.
A sediment from the juice of the fruit of Ecballium Elaterium. Light friable, greenish cakes, about 1/10 inch thick; odour faint, tea-like; taste bitter and acrid.

[1]Elaterinum.  Elaterin.  Dose, 1/40-1/10 gr.; 2-6 mgms.
The active principle of elaterium.  Small scales, with a bitter taste. Insoluble in water, or glycerin, soluble 1 in 160 of cold alcohol, readliy in hot; or in chloroform or solutions of the alkalies.

[2]Pulvis Elaterini Compositus.  Dose, 1-4 grs.; ½-2½ dgms. —2.5; Milk Sugar, 97.5.

*Emplastra*.  (see Ammoniacum, Belladonna, Cantharis, Hydrargyrum, Menthol, Opium, Pix, Plumbum, Resina, Sapo).

**ERGOTA.  ERGOT.**  Dose, 20-60 grs.; 12-40 dgms.
The dried sclerotium of the fungus, Claviceps purpurea, originating in the ovary of Secale cereale, the rye.  Roughly cylindrical, dark, violet-black grains, with tapering ends, from 1/3-1½ inch in length; pinky white within; odour peculiar; taste disagreeable.  Deteriorates rapidly on keeping especially if not kept absolutely dry.  Active principles are the alkaloid ergotoxine and certain amido-acids e. g. parahydroxyphenylethylamine (see p. 106).

[1]Extractum Ergotæ.  (Ergotin.)  Dose, 2-8 grs.; 1-5 dgms.
A soft alcoholic extract.

[1]Injectio Ergotæ Hypodermica.  Dose, 3-10 min.; 0.2-0.6 c.c. —30; Phenol, 0.9; Water to 100.  3 grs. in 10 min.

[1]Extractum Ergotæ Liquidum.  Dose, 10-30 min.; 0.6-2 c.c.
An aqueous extract with alcohol added.

[3]Infusum Ergotæ.  Dose, 1-2 fl. oz.; 30-60 c.c. —5; boiling Water 100.

[3]Tinctura Ergotæ Ammoniata.  Dose, 30-60 min.; 2-4 c.c. —25; Solution of Ammonia, 10; Alcohol 60% to 100: by percolation.

EUCALYPTI GUMMI.  Dose, 2-5 grs.; 1-3 dgms.
A ruby-coloured exudation from the bark of Eucalyptus rostrata, and probably other species.  In grains or small masses; thin fragments are transparent and of a ruby or garnet-red colour; taste astringent, and tinges the saliva red.  About 80-90% soluble in water, almost entirely soluble in alcohol.

[3]Trochiscus Eucalypti Gummi.  1 gr. with the Fruit Basis.

OLEUM EUCALYPTI.  Dose, ½-3 min.; 0.03-0.2 c.c.
The oil distilled from the fresh leaves of Eucalyptus Globulus, and other species.  Colourless or pale yellow, with an aromatic camphoraceous odour, and a pungent taste, leaving a sensation of coldness in the mouth.

[2]Unguentum Eucalypti.—10 by weight; Hard Paraffin, 40; Soft Paraffin, 50.

*Euonymi Cortex. Euonymus Bark.* The dried root bark of Euonymus atropurpureus. In quilled or curved pieces, 1/12-1/6 inch thick; the outer layer, light ash-grey in colour, soft and friable; the inner surface tawny white and smooth; odour faint but characteristic, taste mucilaginous, slightly acid and bitter.

[1]Extractum Euonymi Siccum. Dose, 1-2 grs.; 6-12 cgms.
An alcoholic extract dried and mixed with calcium phosphate.

**Extracta** (the following; with a dose of ¼-1 gr. Belladonnæ Viride, Belladonnæ Alc., Cannabis Indicæ, Colchici, Nucis Vomicæ, Opii, Physostigmatis, Strammonii, Strophanthi: with a dose of 2-8 grs. Anthemidis, Cascaræ Sagradæ, Colocynthidis Comp., Ergotæ, Gentianæ, Hyoscyami Viride, Jalapæ, Rhei: with a dose of 1-2 grs. Euonymi Siccum: with a dose of 1-4 grs. Aloes Barbadensis; with a dose of 5-15 grs. Krameriæ, Taraxaci: in any quantity Glycyrrhizæ).

**Extracta Liquida** (the following; with a dose of ½-2-20 min. Ipecacuanhæ; with dose 1-3 min. Nucis Vomicæ; with dose 5-15 min. Cinchonæ, Hamamelidis, Hydrastis, Jaborandi; with dose, 5-30 min. Cimicifugæ, Opii; with dose, 10-30 min. Ergotæ; with dose, 45-90 min. Filieis; with dose, 30-60 min. Cascaræ Sagradæ, Cocæ, Glycyrrhizæ; with dose ½-2 fl. dr. Pareiræ, Taraxaci; with dose, 2-4 fl. dr. Sarsæ; without dose, Belladonnæ).

*Fel Bovinum Purificatum. Purified Ox Bile.* Dose, 5-15 grs.; 3-10 dgms.
Evaporated ox bile purified by precipitation with alcohol. A yellowish-green, hydroscopic substance, with a bitter-sweet taste. Soluble in water and in alcohol.

**FERRUM. IRON.** Annealed iron wire or wrought iron nails.

Incompatibles of ferric salts, in general, alkali hydrates and carbonates, (precipitate ferric hydrate, in part prevented by sugar, glycerin, citrates, and tartrates); carbonates of the alkaline earths, borax, alkali phosphates and sulphides; alkali hypophosphites in a neutral solution; iodides in an acid solution; arsenites, tannic acid, benzoates; a change in colour is given with tannic and gallic acids, acetates, salicylates, phenol, acetanilid, antipyrine, phenacetin, many oils, oleoresins, and balsams, morphine. (These colour reactions in some cases occur with the chloride only, and are in all cases more marked with it.) Acacia is gelatinised and albumin precipitated

Incompatibles of ferrous salts, readily oxidised by air, alkali hydrates and carbonates, phosphates, borax, tannic and gallic acids, oxidising reagents.

[3]Ferri Arsenas.   Dose, 1/16-1/4 grs.; 4-16 mgms.
Ferrous Arsenate, Fe$_3$(AsO$_4$)$_2$, 6H$_2$O. mixed with some ferric arsenate and oxide. An amorphous, tasteless, greenish powder. Insoluble in water, readily in hydrochloric acid.

[3]Ferric Carbonas Saccharatus.   Dose, 10-30 grs.; ½-2 gms.
Ferrous Oxycarbonate, xFeCo$_3$, yFe(OH)$_2$, more or less oxidised and mixed with sugar.   Brownish-grey lumps or powder, with a sweetish chalybeate taste, of which Iron Carbonate forms about 1/3.   Only partly soluble in water, soluble in hydrochloric acid.

[3]Ferri Phosphas.   Dose, 5-10 grs. 3-6 dgms.
Ferrous phosphate Fe$_3$(PO$_4$)$_2$, 8H$_2$O, (47%) mixed with ferric phosphate and oxide.   A slate-blue amorphous powder.   Insoluble in water, soluble in hydrochloric acid.

[1]Ferri Sulphas.   Dose, 1-5 grs.; ½-3 dgms.
Ferrous sulphate, FeSO$_4$, 7H$_2$O. Pale blue-green crystals with an astringent taste.  Soluble 1 in 1½ of water, insoluble in alcohol.

[1]Mistura Ferri Composita. (Griffith's Mixture.) Dose, ½-1 fl. oz. 15-30 c.c.
—0.57; Potassium Carbonate, 0.686; Myrrh, 1.37; Sugar, 1.37; Spirit of Nutmeg, 1.04; Rose Water, to 100. A dark-green mixture containing a precipitate of ferrous carbonate.

[1]Ferri Sulphas Exsiccatus.   Dose, ½-3 grs.; 1/4-2 dgms.
Ferrous sulphate from which six molecules of water have been removed by heat.   A white powder slowly soluble in a little more than 2 parts of water.

[1]Pilula Ferri. (Blaud's Pill.) Dose, 5-15 grs.; 3-10 dgms. (In 1-3 pills).
—30; Exsiccated Sodium Carbonate, 19; Gum Acacia, 10; Tragacanth, 3; Syrup, 30; Glycerin, 2; Water q.s. Each pill contains about 1 gr. of Ferrous Carbonate.

[2]Ferrum Redactum.   Reduced Iron.   Dose, 1-5 grs.; ½-3 dgms.
A fine greyish-powder, strongly attracted by the magnet, producing black streaks if rubbed in the mortar. Contains at least 75% of iron, the rest being oxide.  Incompatibles, salts of lead, silver, copper, bismuth, mercury and antimony: may explode if rubbed with potassium permanganate and chlorate.

[3]Trochiscus Ferri Redacti.   1 gr. with the Simple Basis.

[1]Ferrum Tartaratum.   Dose, 5-10 grs.; 3-6 dgms.
Garnet scales sweetish and astringent. Slowly soluble 1 in 1 of water, sparingly in alcohol.

[1]Ferri et Ammonii Citras.   Dose, 5-10 grs.; 3-6 dgms.
A mixture of ferric citrate and ammonium citrate.  Deep red, trans-

parent scales, slightly sweetish and astringent in taste. Soluble 2 in 1 of water, giving a slightly acid solution; almost insoluble in alcohol.

[2]Vinum Ferri Citratis.   Dose, 1-4 fl. dr.; 4-16 c.c.
—1.83; Orange Wine, 100.   8 grs. in 1 fl. oz.

[1]Ferri et Quininæ Citras.   Dose, 5-10 grs.; 3-6 dgrs.
Contains ferric and quinine citrate. Greenish-golden scales, somewhat deliquescent, bitter and chalybeate in taste. Soluble, 1 in 2 of water, the solution being very slightly acid. Contains 1 of quinine in 6.66.

[2]Liquor Ferri Acetatis.   Dose, 5-15 min.; 0.3-1 c.c.
A solution containing ferric acetate. Deep red, with a sour, astringent taste, and an acetous odour. Miscible with water and alcohol in all proportions.

[2]Liquor Ferri Perchloridi Fortis. Made by dissolving Iron Wire in acids. An orange-brown liquid with a strong astringent taste, acid in reaction. Miscible with water and alcohol in all proportions. 22.5 grs. of Ferric Chloride in 110 min.

[1]Liquor Ferri Perchloridi.   Dose, 5-15 min.; 0.3-1 c.c.
—25; Water, 75.   5.5 grs. Ferric Chloride in 110 min.

[1]Tinctura Ferri Perchloridi.   Dose, 5-15 min.; 0.3-1 c.c.
—25; Alcohol, 25; Water, 50.

[3]Liquor Ferri Pernitratis.   Dose, 5-15 min.; 0.3-1 c.c.
An acid solution containing ferric nitrate. Reddish-brown, acid and astringent in taste. Contains 3.3 grs. Ferric Nitrate in 110 min.

[3]Liquor Ferri Persulphatis. A solution of ferric sulphate. Dark red in colour.

[1]Syrupus Ferri Phosphatis.   Dose, 30-60 min.; 2-4 c.c.
Contains 1 gr. of anhydrous ferrous phosphate in 1 fl. dr. Acid in reaction.

[1]Syrupus Ferri Phosphatis cum Quinina et Strychnina. Dose, 30-60 min.; 2-4 c.c.
Acid in reaction. 1 fl. dr. contains 1 gr. of anhydrous ferrous phosphate, 4/5 gr. of quinine sulphate, and 1/32 gr. of strychnine.

[1]Syrupus Ferri Iodidi.   Dose, 30-60 min.; 2-4 c.c.
Contains 1 gr. of ferrous iodide in 11 min.

[2]Vinum Ferri.   Dose, 1-4 fl. dr.; 4-16 c.c.
Iron Wire digested in Sherry for 30 days.

Vinum Ferri Citratis (see Ferri et Ammonii Citratis).

*Ficus. Figs.*   The dried fleshy receptacles of Ficus Carica.

**Filix Mas. Male Fern.** The dried rhizome of Aspidium Filix-mas. 3-6 inches long, 3/4-1 inch in diameter, entirely covered with the hard, persistent, curved, angular, dark-brown bases of the petioles; brown externally, green internally: odour feeble but disagreeable; taste at first sweetish and astringent, but later bitter and nauseous.

    [1]Extractum Filicis Liquidum.  Dose, 45-90 min.; 3-6 c.c.
An ethereal extract containing much oil from which the ether has been evaporated.

    *Fœniculi Fructus.*  *Fennel Fruit.*  The dried ripe fruit of Fœniculum capillaceum. Oblong, more or less curved, 1/5-2/5 inch long and 1/10 inch in diameter; odour aromatic; taste aromatic, agreeable, and sweet. Contains an oil.

    Aqua Fœniculi.—10; Water 100: by distillation.

    *Galbanum.*  *Galbanum.*  Dose, 5-15 gr.; 3-10 dgms.
A gum resin obtained from Ferula galbaniflua. In tears or masses of tears, which are rounded or irregular in form, larger or smaller in size than a pea, yellowish-brown in colour and often dirty, internally opaque and yellowish-white: hard and brittle if cold; becoming sticky and ductile if held in the hand: taste bitter and unpleasant, odour characteristic.

    Pilula Galbani Composita.  Dose, 4-8 grs.; 2½-5 dgms.  (In 1 or 2 pills).
Asafetida, Galbanum and Myrrh, of each 28.5; Syrup of Glucose, q.s.

    GALLA.  GALLS.  Excresences on Quercus infectoria resulting from the puncture and deposit of eggs by Cynips Gallæ tinctoriæ.

    [1]Unguentum Gallæ.—20; Benzoated Lard, 80.

    [2]Unguentum Gallæ cum Opio—92.5; Opium, 7.5.

    *Gelatinum.*  *Gelatin.*  The air-dried product of the action of boiling water on such animal products as skin, tendons, ligaments, and bones. Insoluble in alcohol and ether, soluble in acetic acid. A 2% solution in water should gelatinise on cooling.

    GELSEMII RADIX.  GELSEMIUM ROOT.  The dried rhizome and roots of Gelsemium nitidum. Nearly cylindrical pieces 6 inches or more in length. 1/4-3/4 inches in diameter, brown or dark violet -brown externally: taste bitter, odour slightly aromatic. Active principle gelseminine, an alkaloid.

    [1]Tinctura Gelsemii.  Dose, 5-15 min.; 0.3-1 c.c.
—10; Alcohol 100: by percolation.

    **Gentianæ Radix.**  **Gentian Root.**  The dried rhizome and roots of Gentiana lutea. Cylindrical pieces, often longitudinally split, varying in length, but seldom more than an inch in thickness, yellowish-brown externally, reddish-yellow internally: rough from longitudinal wrinkles, and closely approximated, encircling leaf-scars: odour characteristic, taste at first slightly sweet, but afterwards bitter.

[1]Extractum Gentianæ. Dose, 2-8 grs.; 1-5 dgms.
An extract made with hot water and evaporated to the consistence of a soft extract.

[1]Infusum Gentianæ Compositum. Dose, ½-1 fl. oz.; 15-30 c.c. —1.25; Dried Bitter Orange Peel, 1.25; Fresh Lemon Peel, 2.5; boiling Water, 100.

[1]Tinctura Gentianæ Composita. 30-60 min.; 2-4 c.c. —10: Dried Bitter Orange Peel, 3.75; Cardamom Seeds, 1.25; Alcohol 45%, 100; by maceration.

Glucose (See Syrupus Glucosi, p. 89).

*Glusidum. Gluside. (Saccharin.)* Benzoyl sulphonimide,

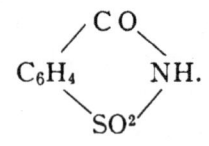

$$C_6H_4 \begin{array}{c} CO \\ \diagdown \\ \diagup \\ SO_2 \end{array} NH.$$

a light, white, crystalline powder, with an intensely sweet taste in dilute solutions.

GLYCERINUM. GLYCERIN. Dose, 1-2 fl. dr.; 4-8 c.c.
$C_3H_5(OH)_3$. Glycerol with a small percentage of water. A clear, colourless syrupy liquid, with a sweet taste; inodorous. Miscible with water and alcohol in all proportions, insoluble in ether, chloroform, and fixed oils.

[1]Suppositoria Glycerini.—70; Gelatin, 14; Water, q.s.

*Glycerina.* (see Boron, Acidum Carbolicum, Acidum Tannicum, Alumen, Amylum, Pepsin, Plumbum, Tragacantha.)

GLYCYRRHIZÆ RADIX. LIQUORICE ROOT. The peeled root and subterranean stem of Glycyrrhiza glabra. In long cylindrical pieces, when peeled, yellow, with a nearly smooth, fibrous surface; odour faint; taste sweet and characteristic. It contains a glucoside, glycyrrhizin, which is its chief sweet principle and is present as a calcium salt. The acid glucoside is insoluble in water and hence is precipitated by acids.

[1]Extractum Glycyrrhizæ. A soft aqueous extract.

[1]Extractum Glycyrrhizæ Liquidum. Dose, 30-60 min.; 2-4 c.c.
An aqueous extract to which alcohol is added.

[3]Pulvis Glycyrrhizæ Compositus. Dose, 60-120 grs.; 4-8 gms. —16; Senna 16; Fennel 8; Sublimed Sulphur, 8; Sugar, 48. 10 grs. Senna and 5 grs. Sulphur in 60 grs. powder.

*Gossypium. Cotton.* The hairs of the seeds of Gossypium Barbadense.

GRANATI CORTEX. POMEGRANATE BARK. The dried bark of the stem and root of Punica Granatum. Usually in curved irregular or channeled pieces, 2-4 inches long, ½-1 inch wide: the outer surface of the root bark rough yellowish-grey, the stem bark smoother, the inner surface is yellow tinged with brown: odourless: taste astringent and bitter. The important active principle is the alkaloid, pelletierine.

Decoctum Granati Corticis. Dose, ½-2 fl. oz.; 15-30 c.c. —20; boiled in 100 of Water.

*Guaiaci Lignum. Guaiacum Wood.* The heart wood of Guaiacum officinale or sanctum. Dark greenish-brown, dense, heavier than water; odour when heated aromatic; taste acrid.

GUAIACI RESINA. GUAIACUM RESIN. Dose, 5-15 grs.; 3-10 dgms. The resin obtained from the bark of Guaiacum officinale. Usually in arge masses but sometimes in tears; brittle in thin splinters transparent, varying in colour from yellowish-green to reddish-brown; odour, more apparent when warmed, balsamic; taste slightly acrid.

Incompatibles, a change in colour to blue is induced in alcoholic solutions by nitric acid, potassium permanganate, ferric chloride, spirit of nitrous ether, and other oxidizing agents; sulphuric acid turns it reddish, and mucilage of acacia blue.

[3]Mistura Guaiaci. Dose, ½-1 fl. oz.; 15-30 c.c. —2.5; Sugar, 2.5; Tragacanth, 0.4; Cinnamon Water, 100.

[1]Tinctura Guaiaci Ammoniata. Dose, 30-60 min.; 2-4 c.c. —20; Oil of Nutmeg, 0.31; Oil of Lemon, 0.21; Strong Solution of Ammonia, 7.5; Alcohol to 100.

[2]Trochiscus Guaiaci Resinæ. 3 grs. with the Fruit Basis.

*Haematoxyli Lignum. Logwood.* The heart wood of Haematoxylon campechianum. Hard, heavy, dull orange to purplish-red externally, internally reddish-brown; odour slight and agreeable; taste sweetish and as tringent. Contains tannic acid and a colouring matter, haematoxylin Incompatibles, mineral acids, metallic salts, especially ferric, lead and antimony, lime-water.

Decoctum Haematoxyli. Dose, ½-1 fl. oz.; 15-30 c.c. —5; Cinnamon Bark, 0.8; boiled with water, and made up to 100.

*Hamamelidis Cortex. Hamamelis Bark.* (Witch Hazel Bark.) The dried bark of Hamamelis virginiana. Usually in curved pieces 28- inches long, 1/16 inch thick; outer surface silvery-grey if covered with the cork, but if freed from it nearly smooth and reddish-brown, the inner surface pale pink, with fine longitudinal striæ: no marked odour; astringent taste.

*Tinctura Hamamelidis.* Dose, 30-60 min.; 2-4 c.c.
—10; Alcohol 45%, 100: by percolation.

*Hamamelidis Folia.* The leaves fresh and dried of Hamamelis virginiana. Broadly oval in outline, 3-6 inches long; upper surface darkgreen to brownish, the lower paler in colour; no marked odour; an astringent, slightly bitter taste.

*Extractum Hamamelidis Liquidum.* Dose, 5-15 min.; 0.3-1 c.c. An alcohol extract.

*Unguentum Hamamelidis.*—10; Hydrous Wool Fat, 90.

*Liquor Hamamelidis.* An alcoholic solution made from the fresh leaves by maceration in alcohol and distillation.

*Hemidesmi Radix. Hemidesmus Root.* The dried root of Hemidesimus Indicus. The root is long, nearly cylindrical, tortuous, and longtudinally furrowed; about 1/4 inch thick; brownish in colour: odour fragrant; taste sweet.

*Syrupus Hemidesmi.* Dose, 30-60 min.; 2-4 c.c.
—10; Sugar, 70; Water, 50.

HIRUDO. LEECHES. Sanguisuga medicinalis and S officinalis.

**Homatropinæ Hydrobromidum. Homatropine Hydrobromide.** Dose, 1/80-1/20 gr.; 3/4-3 mgms.
A white crystalline powder or aggregation of crystals. Soluble 1 in 6 of water, 1 in 18 of alcohol. Incompatibles, as for alkaloids.

[1]*Lamellæ Homatropinæ.* Gelatin disks each containing 1/100 gr.

## HYDRARGYRUM. MERCURY.

Incompatibility, most salts of mercury are insoluble and hence the range of incompatibility of the soluble salts is a wide one: amongst the substances producing precipitation in solution of mercuric salts are alkali hydrates or carbonates including ammonium, lime-water, borax, soluble iodides and bromides (precipitate soluble in excess), phosphates, hypophosphites, and sulphites, arsenites, ferrous salts, tartarated antimony, tannic acid, albumin, gelatin, some bitter principles, and glucosides. With mercurous salts the reactions are similar with the addition that iodides lead to the formation of metallic mercury and mercuric iodide, the same is true of chlorides; oxidising reagents lead to the formation of mercuric salts; cane-sugar, milk-sugar, acacia and tragacanth reduce mercurous salts.

[1]*Hydrargyrum.* A silvery white metal. Volatilises with heat.

[1]Hydrargyrum cum Creta. (Grey Powder.) Dose, 1-5 grs.; ½-3 dgms.
—33; Prepared Chalk, 66.

[3]Emplastrum Hydrargyri. Mercurial Plaster.—32.8; Olive Oil, 1.4; Sublimed Sulphur, 0.2; Lead Plaster, 65.6

[1]Pilula Hydrargyri. (Blue Pill.) Dose, 4-8 grs.; 2½-5 dgms. (in 1 or 2 pills).
—33; Confection of Roses, 49.5; Liquorice Root, 16.5. 1 gr. mercury in 3 grs.

[3]Emplastrum Ammoniaci cum Hydrargyro
—41. Ammoniacum 164, Olive Oil 1.75, Sublimed Sulphur 0.25.

[1]Unguentum Hydrargyri.—48; Lard, 48; Prepared Suet, 3.

[3]Linimentum Hydrargyri.—33; Strong Solution of Ammonia, 11; Liniment of Camphor q.s. (about 55).

[3]Unguentum Hydrargyri Compositum. (Scott's Dressing.) —40: Yellow Beeswax, 24; Olive Oil (by weight), 24; Camphor, 12.

[1]Hydrargyri Iodidum Rubrum. Dose, 1/32-1/16 gr.; 2-4 mgms. Mercuric iodide (biniodide). A vermillion crystalline powder. Almost insoluble in water, sparingly in alcohol, freely in ether, and in solutions of potassium iodide.

[3]Unguentum Hydrargyri Iodidi Rubri.—4; Benzoated Lard, 96.

[3]Hydrargyri Oleas. Mercuric Oleate. An unctuous substance of a light greyish yellow colour, liable to darken on keeping.

[3]Unguentum Hydrargyri Oleatis.—25; Benzoated Lard, 75·

[2]Hydrargyri Oxidum Flavum. Yellow Mercuric Oxide. A yellow non-crystalline powder, insoluble in water.

[2]Unguentum Hydrargyri Oxidi Flavi.—2; Soft Paraffin, yellow, 98.

[1]Hydrargyri Oxidum. Red Mercuric Oxide. Orange-red crystalline scales or powder.

[1]Unguentum Hydrargyri Oxidi Rubri.—10; Paraffin Ointment, yellow, 90.

[1]Hydrargyri Perchloridum. Mercuric Chloride. (Bichloride, Perchloride, Corrosive Sublimate.) Dose, 1/32-1/16 gr.; 2-4 mgms. $HgCl_2$. Heavy colourless crystalline masses, with a highly acrid metallic taste. Soluble 1 in 19 of cold, 1 in 2 of boiling water; 1 in 5 of alcohol; 1 in 4 of ether; 1 in 2 of glycerin with trituration.

[1]Liquor Hydrargyri Perchloridi. Dose, 30-60 min.; 2-4 c.c. —0.114; Water, 100.

[1]Hydrargyri Subchloridum. Mercurous Chloride. (Subchloride, Calomel.) Dose, ½-5 grs.; 1/4-3 dgms.

Hg₂Cl₂. A dull white, heavy, nearly tasteless powder. Insoluble in water, alcohol or ether.

²Pilula Hydrargyri Subchloridi Composita. (Compound Calomel Pill, Plummer's Pill.) Dose, 4-8 grs.; 2½-4 dgms. (In 1 or 2 pills). —22.5; Sulphurated Antimony, 22.5; Guaiacum Resin, 45. Castor Oil (by weight), 9.27; Alcohol, q.s. Almost 1 gr. calomel in 4 grs.

²Unguentum Hydrargyri Subchloridi.—10; Benzoated Lard, 90.

²Hydrargyrum Ammoniatum. Ammoniated Mercury. A white powder, but little acted upon by water.

²Unguentum Hydrargyri Ammoniati.—10; White Paraffin Ointment, 90

³Liquor Hydrargyri Nitratis Acidus. Mercuric nitrate in solution in nitric acid.

¹Unguentum Hydrargyri Nitratis. (Citrine Ointment.) Mercury, 4; in Nitric Acid, 12. Cooled and added to a mixture of Lard, 16 in Olive Oil, 28.

²Unguentum Hydrargyri Nitratis Dilutum.—25; Soft Paraffin, 75.

¹Lotio Hydrargyri Nigra. Black Mercurial Solution. (Black Wash.) Mercurous Chloride, 0.685; Glycerin, 5; Mucilage of Tragacanth. 12.5; Solution of Lime, to 100. The Black oxide is formed $Hg_2Cl_2 + Ca(OH)_2 = Hg_2O + CaCl_2 + H_2O$.

¹Lotio Hydrargyri Flava. Yellow Mercurial Lotion. (Yellow Wash.) Mercuric Chloride, 0.46; Solution of Lime to 100. The yellow oxide is formed. $HgCl_2 + Ca(OH)_2 = HgO + CaCl_2 + H_2O$.

¹Liquor Arsenii et Hydrargyri Iodidi. Dose, 5-20 min.; 0.3-1.2 c.c. 1 of Arsenious Iodide and 1 of Mercuric Iodide in 100 of Water.

HYDRASTIS RHIZOMA. HYDRASTIS RHIZOME. (Golden Seal.) The dried rhizome and roots of Hydrastis Canadensis. The rhizome is tortuous, often branched, ½-1½ inch long, 1/8-½ inch thick; yellowish-brown, on the upper surface ascending branches and on the lower numerous thin brittle roots: slight but characteristic odour; taste bitter. Active principles the alkaloids, hydrastine and berberine.

¹Extractum Hydrastis Liquidum. Dose, 5-15 min.; 0.3-1 c.c. An alcoholic extract.

²Tinctura Hydrastis. Dose, 30-60 min.; 2-4 c.c. —10; Alcohol 60%, 100: by percolation.

Hydrogenii Peroxidi (See Liquor Hydrogenii Peroxidi, p. 71.)

**Hyoscyami Folia. Hyoscyamus** (Henbane) **Leaves.** The fresh leaves, flowers, and branches of Hyoscyamus niger, also the same dried. The leaves vary in length but are seldom more than 10 inches long; oblong, with a conspicuous mid-rib, pale green, with hairs especially along the veins and on the under surface. The flower yellowish, with purplish veins. The herb has a strong characteristic odour, and a bitter slightly acrid taste. Active principles the alkaloids hyoscine and hyoscyamine.

[1]Extractum Hyoscyami Viride. Dose, 2-8 grs.; 1-5 dgms.
The juice expressed from the fresh herb evaporated to a soft extract.

[1]Pilula Colocynthidis et Hyoscyami (see Colocynthis, 54).

[3]Succus Hyoscyami. Dose, 30-60 min.; 2-4 c.c.
Fresh expressed juice, 75; Alcohol, 25.

[1]Tinctura Hyoscyami. Dose, 30-60 min.; 2-4 c.c.
Dried leaves and tops, 10; Alcohol, 45%, 100: by percolation.

**Hyoscinæ Hydrobromidum.** (Scopolamine Hydrobromide.) Dose 1/200-1/100 gr.; 1/3-2/3 mgms.
This alkaloid is also obtained from Datura alba and Scopola. Colourless, transparent crystals, odourless,; taste acrid and slightly bitter. Soluble 1 in 2 of water, 1 in 13 of alcohol. Incompatibles, as for alkaloids but is not precipitated by bicarbonates or ammonium carbonate; decomposed by alkalies or water if warmed.

**Hyoscyaminæ Sulphas.** Dose, 1/200-1/100 gr.; 1/3-2/3 mgms.
A crystalline powder, odourless, deliquescent, with a bitter, acrid taste. Soluble 1 in ½ of water, 1 in 4.5 of alcohol, very slightly in ether or chloroform. Incompatibles as for hyoscine.

*Infusæ.* (The following with a dose of ½-1 fl. oz. Aurantii, Aurantii Comp., Calumbæ, Caryophylli, Cascarillæ, Chiratæ, Cinchonæ Acidum, Gentianæ Comp., Quassiæ, Rhei, Rosæ Acidum, Scoparii, Senegæ, Serpentariæ, Uvæ Ursi; with a dose of ½-2 fl. oz. Sennæ; with a dose of 1-2 fl. oz. Buchu, Cuspariæ, Ergotæ, Krameriæ, Lupuli; with a dose of 2-4 fl. dr. only, Digitalis.)

*Injectiones Hypodermicæ.* (The following with a dose of 2-5 min. Cocainæ, Morphinæ; with a dose of 3-10 min. Ergotæ; with a dose of 5-10 min. Apomorphinæ.)

**Iodoformum. Iodoform.** Dose, ½-3 grs.; ¼-2 dgms.
Tri-iodomethane, $CHI_3$. Shining lemon-yellow crystals, somewhat unctous to the touch, with a persistent characteristic and disagreeable odour and taste. Very sparingly soluble in water or benzol, more soluble in alcohol 1 in 120, ether 1 in 7, chloroform 1 in 14, glycerin 1 in 100, olive oil 1 in 30, and in other fixed oils and lanolin.

[2]Suppositoria Iodoformi. Each suppository contains 3 gr. —20; Oil of Theobroma, to 100.

[2]Unguentum Iodoformi.—10; Paraffin Ointment, to 100.

**Iodum. Iodine.** In crystals, of a dark colour and metallic lustre, yielding if heated, violet fumes. Soluble 1 in 5,000 of water, readily in ether, alcohol, chloroform, or a solution of potassium iodide.

Incompatibles, alkali hydrates or carbonates, ammonia, nitric acid, hypophosphites, sulphites, chlorates, mercurous salts,; in the presence of alkalies, metallic iron, ferrous and arsenous salts; lime-water, tannic acid, fixed oils, volatile oils especially turpentine, alkaloids.

[1]Liquor Iodi Fortis. (Linimentum Iodi.)— 12; Potassium Iodide, 7.2; Water, 12; Alcohol, 86.4. Approximately 1 in 10.

[1]Tinctura Iodi. Dose, 2-5 min.; 0.1-0.3 c.c. —2.5 Potassium Iodide, 2.5; Water, 2.5; Alcohol to 100, approximately equal to 4.4 I or 5.8 KI.

[1]Unguentum Iodi.—4; Potassium Iodide, 4; Glycerin, 12; Lard, 80.

**Ipecacuanhæ Radix. Ipecacuanha Root.** Dose, as an expectorant, 1/4-2 grs.; 15-120 mgms.: as an emetic, 15-30 grs.; 1-2 gms.
The dried root of Psychotria Ipecacuanha. Somewhat tortuous pieces, rarely longer than 6 inches, or thinner than 1/4 inch, in colour varying from dark-red to dark red-brown; odour slight, taste bitter. Active principle the alkaloid, emetine.

[2]Extractum Ipecacuanhæ Liquidum. Dose, as an expectorant, ½-2 min.; 0.03-0.12 c.c.; as an emetic, 15-20 min.; 1-12 c.c.
An alcoholic extract containing calcium hydroxide, and standardised to contain 2-2.25% of alkaloids.

[3]Acetum Ipecacuanhæ. Dose, 10-30 min.; 0.6-2 c.c. —5; Diluted Acetic Acid, 85; Alcohol, 10. Contains 1/10% of alkaloids.

[1]Vinum Ipecacuanhæ. Dose, as an expectorant, 10-30 min.; 0.6-2 c.c.: as an emetic, 4-6 fl. dr.: 16-24 c.c.—5, Vinum Xericum, 100.

[1]Pulvis Ipecacuanhæ Compositus. (Dover's Powder.) Dose, 5-15 grs.; 3-10 dgms.
—10; Opium, 10; Potassium Sulphate, 80. 1 gr. of Opium and 1 of Ipecacuanha in 10 grs.

[3]Pilula Ipecacuanhæ cum Scilla. Dose, 4-8 grs.; 2½-5 dgms. (in 1 or 2 pills).
—60; Squill, 20; Ammoniacum, 20; Syrup of Glucose, q.s. In each pill about 1/4 gr. of opium.

[3]Trochiscus Ipecacuanhæ. 1/4 gr. with the Fruit Basis.

[3]Trochiscus Morphinæ et Ipecacuanhæ.—1/12 gr.; Morphine Hydrochloride, 1/36 gr., with the Tolu Basis.

JABORANDI FOLIA. JABORANDI LEAVES. The dried leaflets of Pilocarpus Jaborandi. Dull green oblong, 2½-4 inches long, glabrous or almost so, showing oil-glands if held up to the light, odour aromatic when bruised, taste at first bitter and aromatic afterwards pungent and increasing the flow of saliva. The active principle is the alkaloid pilocarpine.

[2]Extráctum Jaborandi Liquidum. Dose, 5-15 min.; 0.3-1 c.c. Alcoholic.

[3]Tinctura Jaborandi. Dose, 30-60 min.; 2-4 c.c.
—20; Alcohol 45%, 100: by percolation.

**PILOCARPINÆ NITRAS.** Dose, 1/20-½ gr.; 3-30 mgms.
A white crystalline powder. Soluble 1 in 8-9 of water, slightly in cold, freely in hot alcohol.

**Jalapa. Jalap.** Dose, 5-20 grs.; 3-12 dgms.
The dried tubercules of Ipomœa purga. Dark brown, irregularly oblong, ovoid, napiform or fusiform roots, from 1-3 inches long, hard, compact and heavy; externally wrinkled and furrowed and marked with small transverse scars. Internally varying in colour from yellowish-grey to dingy brown. Odour characteristic, taste at first sweet but afterwards acrid and disagreeable. The resin contains the active principles, which are two glucosides, jalapin and scammonin.

[2]Extractum Jalapæ. Dose, 2-8 grs.; 1-5 dgms.
A dried product obtained by evaporation of an alcoholic and an aqueous extract.

[1]Pulvis Jalapæ Compositus. Dose, 20-60 grs.; 1.2-4 gms.
—33.3; Acid Potassium Tartrate, 60; Ginger, 6.6.

[1]Jalapæ Resina. Dose, 2-5 grs.; 1-3 dgms.
Dark brown opaque fragments translucent at the edges, brittle; odour sweetish; taste acrid. Soluble, readily in alcohol, insoluble in water.

[3]Tinctura Jalapæ. Dose, 30-60 min.; 2-4 c.c.
Standardised to contain 1.5% of Jalap Resin.

Juniper (see Oleum Juniperi p. 76).

KAOLINUM. KAOLIN. A native aluminium silicate, powdered and freed from gritty particles by elutriation. A soft whitish powder. Insoluble in water.

KINO. KINO. Dose, in powder, 5-20 grs.; 3-12 dgms.
The juice obtained from the incisions in the bark of Pterocarpus marsupium. In small, angular, brittle, reddish-black fragments; inodourous, very astringent and tinges the saliva red if chewed. Partially soluble in water, almost entirely soluble in alcohol. Contains a tannic acid.

[1]Pulvis Kino Compositus.  Dose, 5-20 grs.; 3-12 dgms.
—75; Opium, 5; Cinnamon Bark, 20.  1 gr. Opium in 20 grs.
[2]Tinctura Kino.  Dose, 30-60 min.: 2-4 c.c.
—10; Glycerin, 15; Water, 25; Alcohol, to 100: by maceration.

KRAMERIÆ RADIX. KRAMERIA ROOT. (Rhatany.) The dried root of Krameria triandra and K. argentea. Both kinds of root have an astringent taste and tinge the saliva red if chewed.

[3]Extractum Krameriæ.  Dose, 5-15 grs.; 3-10 dgms.
A dried aqueous extract.

[2]Trochiscus Krameriæ.  1 gr. in each with the Fruit Basis.

[2]Trochiscus Krameriæ et Cocainæ.  1 gr.; Cocaine Hydrochloride, 1/20 gr. with the Fruit Basis.

[3]Infusum Krameriæ.  Dose, ½-1 fl. oz.; 15-30 c.c.
—5; boiling Water, 100.

[3]Liquor Krameriæ Concentratus.  Dose, 30-60 min.; 2-4 c.c.
—50; Alcohol 20% to 100: by percolation.

[2]Tinctura Krameriæ.  Dose, 30-60 min.; 2-4 c.c.
—20; Alcohol 60%, 100 by percolation.

*Lamellæ* (see Atropina, Cocaina, Homatropina, Physostigmina).

LAUROCERASI FOLIA. CHERRY-LAUREL LEAVES. The fresh leaves of Prunus Laurocerasus. Thick, somewhat oblong, leaves 5-7 inches long, dark-green, smooth and shining above, much paler beneath; inodourous, but emitting when bruised an odour like bitter almonds. Contain a small amount of hydrocyanic acid.

[1]Aqua Laurocerasi.  Dose, ½-2 fl. dr.; 2-8 c.c.
Standardized to contain 0.1% of hydrocyanic acid.

Lavandula (see Oleum Lavandulæ p. 76).

LIMONIS CORTEX. LEMON PEEL. The fresh outer part of the pericarp of the fruit of Citrus Medica, var. Limonum.

[1]Syrupus Limonis.  Dose, 30-60 min.; 2-4 c.c.
—2; Alcohol, q.s.; Lemon Juice, 50; Sugar, 76.

[3]Tinctura Limonis.  Dose, 30-60 min.; 2-4 c.c.
—25; Alcohol, 100: by maceration.

[1]Oleum Limonis.  Dose, ½-3 min.; 0.05-0.2 c.c.
A pale yellow, fragrant oil; taste warm, bitter and aromatic.

*Succus Limonis. Lemon Juice.* The freshly expressed juice of the ripe fruit of Citrus Medica.

LINIMENTI. (Aconiti, Ammoniæ, Belladonnæ, Calcis, Camphoræ, Ammoniatum, Chloroformi, Crotonis, Hydrargyri, Opii, Potassii Iodidi cum Sapone, Saponis, Sinapis, Terebinthinæ. Terebinthinæ Aceticum.)

LINUM. LINSEED. The dried ripe seeds of Linum usitatissumum. Small brown glassy, nearly flat seeds, about 1/6-1/4 inch long; odourless; taste oily and mucilaginous.

[1]Linum Contusum. Crushed Linseed. The above powdered. It should not be rancid.

[2]Oleum Lini. Made by expressing the seeds. Viscid yellow, with a faint odour and a bland taste. Soluble 1 in 10 of alcohol, and in oil of turpentine.

**Liquores.** (With a dose of ½-1 min. Atropinæ Sulphatis,; ½-2 min. Trinitrini; 2-8 min. Arsenicalis, Arsenici Hydrochloricus, Sodii Arsenatis, Strychninæ Hydrochloridi; 5-15 min. Ferri Arsenatis, Ferri Perchloridi, Ferri Persulphatis, Thyroidei; 5-20 min. Arsenii et Hydrargyri Iodidi; 10-20 min. Sodæ Chlorinatæ; 10-30 min. Potassæ; 10-60 min. Morphinæ Acetatis, Morphinæ Hydrochloridi, Morphinæ Tartratis; 20-60 min. Calcis Saccharatus, Ethyl Nitritis; 30-60 min. Bismuthi et Ammonii Citratis, Hydrargyri Perchloridi; ½-2 fl. dr. Hydrogeni Peroxidi; 2-4 fl. dr. Potassii Permanganatis; 2-6 fl. dr. Ammonii Acetatis, Ammonii Citratis; 1-2 fl. oz. Magnesii Carbonatis 1-4 fl. oz. Liquor Calcis; without stated dose, either stock solutions or for external use, Acidi Chromici, Ammoniæ, Ammoniæ Fortis, Calcis Chlorinatæ, Caoutchouc, Epispasticus, Ferri Perchloridi Fortis, Ferri Persulphatis, Hamamelidis, Iodi Fortis, Picis Carbonis, Plumbi Subacetatis Dilutus, Plumbi Subacetatis Fortis, Sodii Ethylatis, Zinci Chloridi.)

*Liquores Concentrati.* (With dose of 30-60 min. Calumbæ, Chiratæ, Cuspariæ, Krameriæ, Quassiæ, Rhei, Senegæ, Sennæ; with dose, ½-2 fl. dr. Serpentariæ; dose, 2-8 fl. dr. Sarsæ Comp.)

*Liquor Ethyl Nitris. Solution of Ethyl Nitrite.* Dose, 20-60 min.; 1.2-4 c.c.
A mixture of 95 parts by weight of absolute alcohol and 5 parts of glycerin, containing between 2½ and 3% of ethyl nitrite. A highly inflammable, limpid, colourless liquid with a characteristic apple-like odour.

LIQUOR HYDROGENII PEROXIDI. SOLUTION OF HYDROGEN PEROXIDE. Dose, ½-2 fl. dr. 2-8 c.c.
An aqueous solution of hydrogen peroxide. A colourless, odourless liquid with a slightly acid taste; renders saliva frothy.

LIQUOR PANCREATIS. PANCREATIC SOLUTION. An alcoholic preparation containing the digestive principles of the pig's pancreas.

*Liquor Sodii Ethylatis. Solution of Sodium Ethylate.* A colourless liquid of syrupy consistence, becoming brown on keeping. Made by dissolving sodium in absolute alcohol.

**Liquor Trinitrini. Solution of Trinitrin.** (Nitroglycerine Glonoin.) Dose, ½-2 min.; 0.03-0.12 c.c.
Trinitroglycerine of commerce, 1; Alcohol, 100. A clear and colourless liquid.

LITHIUM. LITHIUM.
Incompatible with lithium salts in solution are carbonates and phosphates.
[1]Lithii Carbonas. Dose, 2-5 grs.; ½-3 dgms.
A white powder or minute crystalline grains; in solution has an alkaline reaction. Soluble 1 in 70 of water, insoluble in alcohol.
[1]Lithii Citras. Dose, 5-10 grs.; 3-6 dgms.
A white crystalline deliquescent salt. Soluble 1 in 2 of water.
[1]Lithii Citras Effervescens. Dose, 60-120 grs.; 4-8 gms.
—5; Sodium Bicarbonate, 58; Tartaric Acid, 31; Citric Acid, 21. A granular powder.

LOBELIA. LOBELIA. The dried flowering herb of Lobelia inflata. Stems are angular, channelled, and furnished with narrow wings, purplish in colour, hairy. The leaves are irregularly toothed and hairy. Odour somewhat irritant. Taste at first not marked but subsequently burning and acrid.
Tinctura Lobeliæ Ætherea. Dose, 5-15 min.; 0.3-1 c.c.
—20; Spirit of Ether, 100: by percolation.

*Lotiones* (see Hydrargyrum p. 66).

*Lupulus. Hops. (Humulus.).* The dried fruits (strobiles) of Humulus Lupulus. The fruits are about 1¼ inches long, and consist of a number of imbricated bracts and stipules. The odour is aromatic and characteristic; taste bitter, aromatic, and somewhat astringent.
[3]Tinctura Lupuli. Dose, 30-60 min.; 2-4 c.c.
—20; Alcohol 60% 100: by maceration.
[3]Infusum Lupuli. Dose, 1-2 fl. oz.; 15-30 c.c.
—5; boiling Water, 100.
[2]Lupulinum. Lupulin. Dose, 2-5 gr.; 1-3 dgms.
Glands obtained from the fruits of Humulus Lupulus. A granular yellow brown powder, consisting of the minute glands, odour strong and hop-like, taste bitter and aromatic.

**Magnesium. Magnesium.**
Incompatibility, soluble salts of magnesium in strong solutions are precipitated by the hydrates of the alkalies and the alkaline earths; alkali carbonates, phosphates, arsenates; sulphides, oxalates, tartrates.

[2]Magnesia Levis. Light Magnesia. (Light Calcined Magnesia, Light Magnesium Oxide.) Dose, if repeated, 5-30 grs.; 3-20 dgms.: for a single administration, 30-60 grs.; 2-4 gms.

A bulky white powder. Insoluble in water.

[2]Magnesia Ponderosa. Heavy Magnesia. (Heavy Calcined Magnesia, Heavy Magnesium Oxide.) Dose, as above for the Light Magnesia.

A white powder insoluble in water. Differs in weight only from the Light Magnesia, the same weight having only 2/7 of the volume of that of the Light.

[1]Magnesii Carbonas Levis. Light Magnesium Carbonate. Dose as for Light Magnesia.
A very light powder. Insoluble in water.

[1]Magnesii Carbonas Ponderosus. Dose, as for Light Magnesia.
A heavy white powder. Insoluble in water.

[1]Magnesii Sulphas (Epsom Salt). Dose, if repeated, 30-120 grs.; 2-8 gms.: for a single administration 1/4-½ oz.; 8-15 gms.
Small, colourless crystals, with a bitter taste. Soluble, 10 in 13 of water, insoluble in acohol.

[2]Magnesii Sulphas Effervescens. Dose, if repeated, 60-240 grs.; 4-16 gms.: for a single administration, ½-1 oz.; 15-30 gms.
—50; Sodium Bicarbonate, 36; Tartaric Acid,19; Citric Acid, 12.5; Sugar, 10.5.

[3]Liquor Magnesii Carbonatis. Dose, 1-2 fl. oz.; 30-60 c.c. Magnesium Sulphate, 10; Sodium Carbonate, 12.5; Water to 100.

*Mel Depuratum. Clarified Honey.* The honey of commerce melted and strained through flannel.

[2]Oxymel. Oxymel. Dose, 1-2 fl. dr.; 4-8 c.c.
—80; Acetic Acid, 10; Water to 100.

*Mellita* (see Boron p. 42).

**Menthol. Menthol.** Dose, ½-2 grs.; 3-12 cgms.
$C_6H_9.OHCH_3.C_3H_7$. A saturated secondary alcohol obtained from various species of Mentha. Colourless, brittle crystals, with a strong odour of peppermint, and a warm aromatic taste followed by a sensation of cold on drawing air into the mouth. Almost insoluble in water and glycerin,

soluble 5 in 1 of alcohol, 8 in 3 of ether, 4 in 1 of chloroform, 1 in 4 of olive oil, and in other oils.

Incompatibility, when triturated gives a liquid or soft mass with butylchloral, camphor, phenol, chloral, resin, resorcin, thymol.

Emplastrum Menthol.—15; Yellow Wax, 10; Resin, 75.

*Mezerei Cortex. Mezereon Bark.* The dried bark of Daphne Mezereum, D. Laureola, or D. Gnidium. In long thin, more or less flattened strips, or quills of various lengths, flexible, very tough and fibrous; outer surface brown, inner white and silky; no marked odour but an acrid burning taste.

MISTURÆ (see Ammoniacum, Amygdala, Creosotum, Creta, Ferrum, Guaiacum, Oleum Ricini, Senna, Vinum Gallicum). Dose, ½-1 fl. oz.; save that the last three have a dose of 1-2 fl. oz.

Morphina (see Opium p. 79).

*Moschus. Musk.* Dose, 5-10 gr.; 3-6 dgms.
The dried secretions from the preputial follicles of Moschus moschiferus.

*Mucilagines* (see Acacia, Tragacanth).

*Myristica. Nutmeg.* The dried seeds of Myristica fragrans. Oval or rounded seeds rarely exceeding an inch in length; odour strong and agreeably aromatic; taste aromatic, warm and slightly bitter.

Oleum Myristicæ. Dose ½-3 min.; 0.03-0.2 c.c.
A colourless or pale yellow oil having the taste of nutmeg. Soluble 1 in 4½ of alcohol, in all proportions in absolute alcohol.

Spiritus Myristicæ. Dose, 5-20 min.; 0.3-1.2 c.c.
—10; Alcohol, to 100.

MYRRHA. MYRRH. A gum resin obtained from the stem of Balsamodendron Myrrha. Rounded or irregular tears or masses of tears, reddish externally, dry and brittle and more or less covered with a fine powder: odour aromatic, taste aromatic, acrid and bitter: contains 40-60% of gum soluble in water; remainder is largely resin and is soluble in alcohol.

[1]Pilula Aloes et Myrrhæ (see Aloe p. 32).
[1]Tinctura Myrrhæ. Dose, 30-60 min.; 2-4 c.c.
—20; Alcohol, 100: by maceration.

NAPHTHOL BETA-NAPHTHOL. Dose, 3-10 grs.; 2-6 dgms. Betamonohydroxynaphthalene, $C_{10}H_7OH$. White or nearly white crystalline laminæ or in powder; taste sharp and pungent; odour resembling phenol. Soluble 1 in 1,000 of water, 1 in less than 2 of alcohol.

**NUX VOMICA. NUX VOMICA.** Dose, in powder, 1-4 grs.; 6-25 cgms.
The dried ripe seeds of Strychnos Nux-vomica. Nearly disc-shaped, greyish in colour, ¾-1 inch in diameter, and ¼ inch thick; concavoconvex nearly flat, but sometimes irregularly bent; taste bitter.

[1]Extractum Nucis Vomicæ Liquidum. Dose, 1-3 min.; 0.06-0.2 c.c.
An alcoholic extract standardised to contain 1.5 gr. of strychnine in 110 min.

[1]Extractum Nucis Vomicæ. Dose ¼-1 gr.; 15-60 mgms. The liquid extract evaporated and Milk-sugar added, and standardised to contain 5% of strychnine.

[1]Tinctura Nucis Vomicæ. Dose, 5-15 min.; 0.3-1 c.c.
—16.6; Water, 25; Alcohol to 100. Contains ¼ gr. strychnine in 110 min.

**STRYCHNINA. STRYCHNINE.** Dose, 1/60-1/15 gr.; 1-4 mgms. Colourless, inodorous crystals. Soluble 1 in 7,000 of cold, 1 in 2,500 of hot water, 1 in 170 of alcohol, 1 in 6 of chloroform.

Incompatibles as for alkaloids.

[1]Syrupus Ferri Phosphatis cum Quinina et Strychnina. (see p. 60.)

**STRYCHNINÆ HYDROCHLORIDUM.** Dose, 1/60-1/15 gr.; 1-4 mgms. Small colourless crystals, which readily effloresce in the air. Soluble, 1 in 35 of water, 1 in 60 of alcohol.

[1]Liquor Strychninæ Hydrochloridi. Dose, 2-8 min.; 0.1-0.5 c.c.
—1; Alcohol, 25; Water to 100. 1 gr. strychnine in 110 min.

*Olea* (see Amygdala, Anethum, Anisum, Anthemis, Caruum, Caryophyllum, Cinnamomum, Copiaba, Coriandrum, Cubeba, Limon, Linum, Myristica, Phosphorus, Pimenta, Rosa, Sinapis, and the following).

*Oleum Cadinum. Oil of Cade.* (Juniper Tar Oil.) An empyreumatic oily liquid obtained by the destructive distillation of the woody portions of Juniperus Oxycedrus. A dark reddish-brown, almost black, more or less viscid oily liquid, with a not unpleasant empyreumatic odour and an aromatic, bitter and acrid taste. Slightly soluble in water, partially soluble in alcohol.

*Oleum Cajuputi. Oil of Cajuput.* Dose, ½-3 min.; 0.03-0.2 c.c. The oil distilled from the leaves of Melaleuca Leucadendron. Blueish-green, with a penetrating camphoraceous odour, and an aromatic, bitterish taste.

Spiritus Cajuputi. Dose, 5-20 min.; 0.3-1.2 c.c.
—10; Alcohol to 100.

**OLEUM CROTONIS. CROTON OIL.** Dose, ½-1 min.; 0.03-0.06 c.c. The oil expressed from the seeds of Croton Tiglium. Brownish-yellow to dark reddish-brown, viscid, with a disagreeable odour, and an acrid, burning taste. Entirely soluble in absolute alcohol, freely soluble in ether and chloroform.

[3]Linimentum Crotonis.—12.5; Oil of Cajuput, 43.75; Alcohol, 43.75.

OLEUM EUCALYPTI. OIL OF EUCALYPTUS. Dose, ½-3 min.; 0.03-0.2 c.c.
The oil distilled from the fresh leaves of Eucalyptus globulus and other species. Colourless or pale yellow oil, with an aromatic, camphoraceous odour, and a pungent taste; leaves a sensation of coldness in the mouth.

Unguentum Eucalypti.—10 by weight; Hard Paraffin, 40; Soft Paraffin, 50.

*Oleum Juniperi. Oil of Juniper.* Dose, ½-3 min.; 0.03-0.2 c.c.
The oil distilled from the full-grown, unripe, green fruit of Juniperus communis. Colourless or pale green-yellow with the characteristic odour of the fruit and an aromatic, warm, bitterish taste.

Spiritus Juniperi. Dose 20-60 min.; 1.2-4 c.c. —5; Alcohol, 100.

*Oleum Lavandulæ. Oil of Lavender.* Dose, ½-3 min.; 0.03-0.2 c.c.
The oil distilled from the flowers of Lavandula vera. Pale yellow or nearly colourless, with the fragrant odour of the flowers, and a pungent, bitter taste. Soluble 1 in 3 of alcohol 70%.

[3]Spiritus Lavandulæ. Dose, 5-20 min.; 0.3-1.2 c.c. —10; Alcohol, 90.

[1]Tinctura Lavandulæ Composita. Dose, 30-60 min.; 2-4 c.c. —0.47; Oil of Rosemary, 0.05; Cinnamon Bark, 0.85; Nutmeg, 0.85; Red Sanders Wood, 1.7; Alcohol, to 100: by maceration and solution.

**Oleum Menthæ Piperitæ. Oil of Peppermint.** Dose ½-3 min; 0.03-0.2 c.c.
The oil distilled from fresh flowering peppermint, Mentha piperita. Colourless, or pale yellowish, when fresh but darkening with age; odour of peppermint and a strong, aromatic taste, followed by a sensation of coldness in the mouth.

[1]Aqua Menthæ Piperitæ. 1 in about 1,000 by distillation.

[1]Spiritus Menthæ Piperitæ. Dose, 5-20 min.; 0.3-1.2 c.c. —10; Alcohol, 90.

**Oleum Menthæ Viridis. Oil of Spearmint.** Dose, ½-3 min.; 0.03-0.2 c.c.
The oil distilled from the fresh flowering spearmint, Mentha viridis. Colourless, or pale yellowish, when fresh but darkening with age; odour and taste of the herb. Soluble about 1 in 1 of alcohol; soluble in alcohol absolute.

[1]Aqua Menthæ Viridis. 1 in about 1,000, by distillation.

**Oleum Morrhuæ. Cod-liver Oil.** Dose, 1-4 fl. dr.; 4-16 c.c.
The oil extracted from the fresh liver of the cod, Gadus Morrhua. Pale yellow, with a slight fishy, but not rancid, odour. Readily soluble in ether or chloroform and slightly soluble in alcohol.

OLEUM OLIVÆ. Olive Oil. The oil expressed from the ripe fruit of Olea Europæa. A pale yellow oil with a faint odour and a bland taste.

OLEUM PINI. OIL OF PINE. The oil distilled from the fresh leaves of Pinus Pumilio. Colourless or nearly so with a pleasant aromatic odour and pungent taste.

**Oleum Ricini. Castor Oil.** Dose, 1-8 fl. dr.; 4-32 c.c.
The oil expressed from the seeds of Ricinus communis. Viscid almost colourless, almost odourless and a bland taste at first, but afterwards acrid and unpleasant. Soluble 1 in 1 of absolute alcohol, 1 in 5 of alcohol.

[2]Mistura Olei Ricini. Dose, 1-2 fl. oz.; 30-60 c.c.
—37.5; Mucilage of Acacia, 18.75; Orange-flower Water, 12.5 Cinnamon Water, 31.25.

*Oleum Rosmarini. Oil of Rosemary.* Dose, ½-3 min.; 0.03-0.2 c.c.
The oil distilled from the flowering tops of Rosmarinus officinalis. Colourless or pale yellow, with the odour of rosemary, and a warm, camphoraceous taste.

[1]Spiritus Rosmarini.—10; Alcohol to 100.

**Oleum Santali. Oil of Sandal Wood.** Dose, 5-30 min.; 0.3-2 c.c. The oil distilled from the wood of Santalum album. Somewhat viscid, pale yellow oil with a strongly aromatic odour and a pungent, spicy taste.

**Oleum Terebinthinæ. Oil of Turpentine.** Dose, 2-10 min.; 0.1-0.6 c.c.: as an anthelmintic, 3-4 fl. dr.; 12-16 c.c.
The oil distilled, usually with the aid of steam, from the oleo-resin (tur-

pentine) obtained from Pinus sylvestris and other species. Limpid, colourless, with a strong odour, and a pungent and somewhat bitter taste.

[1]Linimentum Terebinthinæ.—65; Soft Soap, 7.5; Camphor, 5.0; Water to 100.

[2]Linimentum Terebinthinæ Aceticum.—44; Glacial Acetic Acid, by weight, 11; Liniment of Camphor, 44.

OLEUM THEOBROMATIS. OIL OF THEOBROMA. (Cacao Butter.) A concrete oil obtained by pressing the warm, crushed seeds of Theobroma Cacao. A yellowish-white solid, with an odour resembling cacao, taste bland and agreeable, free from rancidity. It softens at 80°F. (26.6°C.) and melts between 88°-93° F. (31.1°-33.9° C). Contained in all suppositories except that of Glycerin.

**OPIUM. OPIUM.** Dose, ½-2 grs., 3-12 cgms.
The juice obtained by incision from the unripe capsules of Papaver somniferum, inspissated by spontaneous evaporation. Usually in rounded or more or less irregular large masses, when dry, hard and a dark brown-black colour; odour strong and characteristic, taste bitter. For the use in the preparation of standardised galenical preparations any suitable variety may be used, provided that it contain not less than 7.5% of anhydrous morphine when dry. For all other purposes opium must contain of its dry weight 9.5-10.5% of anhydrous morphine. The chief alkaloidal constituent is morphine, the alkaloids codeine, thebaine, narcotine, are the chief of the other alkaloids that occur. Opium also contains meconic acid, free and in combination. Incompatibles, alkaline carbonates, salts of mercury, and preparations containing tannin; and due to meconic acid, ferric salts (red colour), lead acetate, silver nitrate, barium chloride, calcium chloride, nitric acid.

[3]Emplastrum Opii.—10; Resin Plaster, 80.

[2]Extractum Opii. 1/4-1 gr.; 15-60 mgms.
A partially dried aqueous extract, standardised to contain 20% of morphine.

[1]Extractum Opii Liquidum. Dose, 5-30 min.; 0.3-2 c.c.
Of the extract, 3.75; Alcohol, 20; Water, 80. Contains about 3/4 gr. morphine in 110 min.

[1]Pilula Plumbi cum Opio. Dose, 2-4 grs.; 12-24 cgms. (In 1 or 2 pills).
—12.5; Lead Acetate, 75; Syrup of Glucose, q.s. Each pill contains 1/4 gr. of opium.

[2]Pilula Saponis Composita. Dose, 2-4 grs.; 12-24 cgms. (In 1 or 2 pills.)
—20; Hard Soap, 60; Syrup of Glucose, 20. Each pill contains 2/5 gr. of opium.

[2]Pulvis Opii Compositus. Dose, 2-10 grs.; 1-6 dgms.
—10.5; Black Pepper, 14; Ginger, 35; Caraway Fruit, 42; Tragacanth, 3.5. Contains roughly 1 gr. opium in 10 grs. of powder.

[1]Pulvis Ipecacuanhæ Compositus. (Dover's Powder.) Dose, 5-15 grs.; 3-10 dgms.
—10; Ipecacuanha Root, 10; Potassium Sulphate, 80. 1 gr. opium in 10 grs.

[2]Pilula Ipecacuanhæ cum Scilla. (See Ipecacuanha.) Contains 1 gr. of opium in 20 grs.

[2]Pulvis Kino Compositus. Dose, 5-20 grs.; 3-12 dgms.
—5; Kino, 75; Cinnamon Bark, 20. Contains 1 gr. opium in 20 grs.

[1]Pulvis Cretæ Aromaticus cum Opio. Dose, 10-40 grs.; 6-24 dgms.
—2.5; Aromatic Powder of Chalk, 97.5. Contains 1 gr. of opium in 40 grs.

[3]Suppositoria Plumbi Composita. (See Plumbum.) Contains 1 gr. in each.

[1]Tinctura Opii. (Laudanum.) Dose, if repeated, 5-15 min.; 0.3-1 c.c.: for a single administration, 20-30 min.; 1.2-2 c.c.
A tincture standardised to contain 1 gr. opium in 15 min., or 0.7-0.8% of anhydrous morphine.

[2]Linimentum Opii.—50; Liniment of Soap, 50.

[2]Tinctura Opii Ammoniata. (Scotch Paregoric.) Dose, 30-60 min.; 2-4 c.c.
—15; Benzoic Acid, 2.06; Oil of Anise, 0.625; Solution of Ammonia, 20; Alcohol, to 100. Contains about 0.62 gr. of opium in 1 fl. dr.

[1]Tinctura Camphoræ Composita. (Paregoric, Paregoric Elixir) Dose, 30-60 min.; 2-4 c.c.
—6.09; Benzoic Acid, 0.46; Camphor, 0.34; Oil of Anise, 0.31; Alcohol 60% to 100. Contains 0.25 gr. opium, or 1/30 morphine hydrochloride, in 1 fl. dr.

[2]Unguentum Gallæ cum Opio. Opium, 7.5; Gall Ointment, 92.5.

**MORPHINÆ HYDROCHLORIDUM.** Dose, 1/8-½ gr.; 8-30 mgms.
White acicular, silky crystals or a white crystalline powder. Solubility, 1 in 24 of cold, 1 in 1 of boiling water; 1 in 50 of alcohol.

[1]Liquor Morphinæ Hydrochloridi. Dose, 10-60 min.; ½-4 c.c.
—1; Diluted Hydrochloric Acid, 2; Alcohol, 25; Water, to 100. 1 gr. in 110 min.

[2]Suppositoria Morphinæ. 1/4 gr. in each; Oil of Theobroma, 14.75.

[1]Tinctura Chloroformi et Morphinæ Composita. Dose, 5-15 min.; 0.3-1 c.c.
1/11 gr. of morphine in 10 min. (see Chloroform p. 50).

[2]Trochiscus Morphinæ. 1/36 gr. with the Tolu Basis.

[2]Trochiscus Morphinæ et Ipecuanchæ. 1/36 gr. with 1/12 gr. of Ipecacuanha and the Tolu Basis.

**MORPHINÆ ACETAS.** Dose, 1/8-½ gr.; 8-30 mgms.
A white crystalline or amorphous powder. Soluble 1 in 2½ of water, 1 in 100 of alcohol.

[2]Liquor Morphinæ Acetatis. Dose, 10-60 min.; ½-4 c.c.
—1; Diluted Acetic Acid, 2; Alcohol, 25; Water, to 100.

**MORPHINÆ TARTRAS.** Dose, 1/8-½ gr.; 8-30 mgms.
A white crystalline powder. Soluble, 1 in 11 of water, almost insoluble in alcohol.

[1]Injecto Morphinæ Hypodermica. Dose, 2-5 min.; 0.1—0.3 c.c.
—5; Water 100. 5 gr. of the tartrate in 110 min.

[2]Liquor Morphinæ Tartratis. Dose, 10-60 min.; ½-4 c.c.
—1; Alcohol, 25; Water, to 100.

**Codeina. Codeine.** Dose, 1/4-2 grs.; 15-120 mgms.
An alkaloid obtained from Opium or from morphine. Nearly colourless crystals. Soluble, 1 in 80 of water, 1 in 2 of alcohol.

**Codeinæ Phosphas.** Dose, 1/4-2 grs.; 15-120 mgms.
White bitter crystals. Soluble 1 in 4 of water, 1 in 200 of alcohol.

[2]Syrupus Codeinæ, Dose, ½-2 fl. dr.; 2-8 c.c.
—0.46; Water, 1.25; Syrup, 98.75. 1/4 gr. of codeine phosphate in 1 fl. dr.

*Oxymel* (see Mel and Scilla).

*Papaveris Capsulæ. Poppy Capsules.* The nearly ripe dried fruits of Papaver somniferum.

*Paraffinum Durum. Hard Paraffin.* A mixture of several of the harder paraffins. Colourless, semi-transparent, crystalline, inodourous and tasteless. Melting-point 130°-135° F. (54.4°-57.2° C.). Insoluble in water, slightly soluble in alcohol, readily soluble in ether.

[1]Unguentum Paraffini.—30; Soft Paraffin, 70.

*Paraffinum Liquidum. Liquid Paraffin.* A clear, colourless, odourless, tasteless liquid, obtained from petroleum.

*Paraffinum Molle. Soft Paraffin.* (*Vaseline*). A white or yellow, translucent, soft, unctuous, mixture of the softer members of the paraffin series. Melts at 96°-102° F. (35.5°-38.9° C.). Insoluble in water, slightly soluble in absolute alcohol, readily soluble in ether, chloroform and benzol.

**Paraldehydum. Paraldehyde.** Dose, ½-2 fl. dr.; 2-8 c.c.
A clear colourless liquid, with a characteristic odour and an ethereal, acrid and afterwards cool, taste. Soluble 1 in 10 of water, miscible in alcohol and ether.

*Pareiræ Radix. Pareira. Root.* The dried root of Chondrodendron tomentosum. In long nearly cylindrical and somewhat twisted pieces. from 3/4-2 inches or more in diameter; covered with a thin black bark with numerous furrows, ridges and fissures; internally greyish: no odour, taste bitter.

Extractum Pareiræ Liquidum. Dose, ½-2 fl. dr.; 2-8 c.c. Aqueous and alcoholic.

**Pepsinum. Pepsin.** Dose, 5-10 gr.; 3-6 dgms.
The enzyme obtained from the stomach of the pig, sheep or calf. It should be capable of dissolving 2,500 times its weight of hard-boiled white of egg.

[1]Glycerini Pepsini. Dose, 1-2 fl. dr.; 4-8 c.c.
—9.15; Hydrochloric Acid, 1.15; Glycerin, 60; Water, to 100.

**Phenacetinum. Phenacetin.** Dose, 5-10 grs. 3-6 dgms.
Paraacet-phenetidin. $C_2H_5O.C_6H_4.NHCOCH_3$. White tasteless, inodorous, crystals, neutral to litmus. Soluble, 1 in 1,700 of water, 1 in 21 of alcohol.

**Phenazonum. Phenazone.** (Antipyrine.) Dose, 5-20 grs.; 3-12 dgms.
Phenyldimethyl-iso-pyrazolone. Colourless, inodorous, scaly crystals, with a bitter taste. Soluble, 1 in 1 of water, 1 in a little more than 1 of alcohol. Incompatibles, ferric chloride (gives a red colour), syrup of ferrous iodide, calomel, mercuric chloride, solution of arsenic and mercuric iodides, iodine, potassium permanganate, tannic acid, spirits of nitrous ether and other solutions containing nitrites (a green colour being produced), chloral (in strong solutions); triturated with sodium salicylate a mass or liquid is formed; with thymol, acetananilid, and resorcin a liquid is formed.

**Phosphorus. Phosphorus.** Dose, in pill or solution, 1/100-1/20 gr.; ½-3 mgms.

A semi-transparent, wax-like solid. Insoluble in water, but soluble 1 in 350 of alcohol absolute, 1 in 80 of ether, 1 in 25 of chloroform, 1 in ½ of carbon bisulphide, 1 in 80 of olive oil.

Incompatibles oxidising agents, explodes if triturated with them.

[2]Oleum Phosphoratum. Dose, 1-5 min. 0.06-0.3 c.c.
—1; dissolved at 180° F. in 99 of Almond Oil.

[1]Pilula Phosphori. Dose, 1-2 grs.; 6-12 cgms.
—2; White Beeswax, 25; Lard, 25; Kaolin, 23; Carbon Bisulphide, q.s.; Gum Acacia, q.s. To this mass is to be added 1/3 of its weight of Gum Acacia before dispensing. The pills should be varnished.

**Physostigmatis Semina. Calabar Bean.** The ripe seeds of Physostigma venenosum. Large brownish reniform seeds, usually an inch long by ¾ of an inch wide and ½ inch thick. The active principle is the alkaloid physostigmine.

[3]Extractum Physostigmatis. Dose, ¼-1 gr.; 15-60 mgms. An evaporated alcoholic extract to which three times its weight of Milk-sugar has been added.

**Physostigminæ Sulphas.** (Eserine Sulphate.) Dose, 1/60-1/20 gr.; 1-3 mgms.
Yellowish-white, minute crystals, which turn red on exposure to air and light. Soluble 1 in 4 of water, 2½ in 1 of alcohol.

[1]Lamellæ Physostigminæ. 1/1,000 in gelatin discs.

*Picrotoxinum. Picrotoxin.* Dose, 1/100-1/25 gr.; ½-2½ mgms. Colourless, inodorous, crystals, with a bitter taste. Soluble 1 in 330 of water, 1 in 13 of alcohol.

Pilocarpina (see Jaborandi p. 69).

PILULÆ. (The following with a dose of 1-2 grs. Phosphori; with a dose of 2-4 grs. Plumbi cum Opio, Saponis Comp.; with a dose of 2-8 grs., Quininæ Sulphatis; with a dose of 4-8 grs., Aloes Barbadensis, Aloes et Asafetida, Aloes et Ferri, Aloes et Myrrhæ, Aloes Socotrinæ, Cambogiæ Comp., Colocynthidis Comp., Colocynthidis et Hyoscyami, Galbani Comp., Hydrargyri, Hydrargyri Subchloridi Comp., Ipecacuanhæ cum Scilla, Rhei Comp., Scammonii Comp., Scillæ Comp.; with a dose of 5-15 grs. Ferri.)

*Pimenta. Pimento.* The dried full-grown fruit of Pimenta officinalis. Dark reddish-brown, nearly globular, two-celled, fruits, about 1/5-1/3 inches in diameter; odour and taste characteristic, somewhat like cloves.

Aqua Pimentæ.—5; Water 200: distill off 100.

Oleum Pimentæ. Dose, ½-3 min.; 0.03-0.2 c.c.
Yellowish or yellowish-red oil becoming darker on keeping; odour and taste of pimento.

*Piper Nigrum. Black Pepper.* The dried unripe fruit of Piper nigrum. Almost globular, one-celled fruits, about 1/5 of an inch in diameter; odour aromatic and taste pungent.

Confectio Piperis. Dose, 60-120 grs.; 4-8 gms.
—10; Caraway Fruit 15; Clarified Honey, 75.

*Pix Burgundica. Burgundy Pitch.* The resinous exudation obtained from the stem of Picea excelsa. Hard and brittle yet slowly flowing dull red or yellow-brown; odour aromatic, taste sweet and aromatic.

Emplastrum Picis.—52; Frankincense, 26; Resin, 9; Yellow Beeswax, 9; Olive Oil, 4; Water, 4.

*Pix Carbonis Preparata. Prepared Coal Tar.*
Liquor Picis Carbonatis.—20; Quillaia Bark, 10; Alcohol to 100: by percolation and digestion.

*Pix Liquida. Tar.* A bituminous liquid prepared from the wood of Pinus sylvestris and other species of Pinus, by destructive distillation. A dark brown or blackish semi-liquid substance, of a peculiar aromatic odour.
Soluble 1 in 10 of alcohol.

[2]Unguentum Picis Liquidum.—71.5; Yellow Beeswax, 28.5.

### Plumbum. Lead.

Incompatibles of soluble lead salts, alkali hydrates, carbonates, borax, sulphates, bromides, iodides, phosphates, cyanides, alkali sulphites, benzoates, citrates, tartrates, salicylates, meconates, many colouring matters, resins, glucosides, neutral principles, and alkaloids, in strong solution chlorides.

[1]Plumbi Acetas. (Sugar of Lead). Dose 1-5 grs.; ½-3 dgms.
Pb $(C_2H_3O_2)_2$, $3H_2O$. Small white crystals, slightly efflorescent, with an acetous odour and a sweet astringent taste. Soluble 1 in less than 2 of water, 1 in 20 of alcohol.

Incompatibles, other than the above, with phenol, chloral, salicylic acid, sodium phosphate, gives when triturated a liquid or soft mass with resorcin, and with sodium salicylate a stiff mass.

[1]Pilula Plumbi cum Opio. Dose, 2-4 grs.; 12-24 cgms. (In 1 or 2 pills).
—75; Opium, 12.5; Syrup of Glucose, q.s. ¼ gr. of Opium in each 2 gr. pill.

[3]Suppositoria Plumbi Composita. 1 gr. of opium and 3 grs. of lead acetate in each suppository with Oil of Theobroma.

[3]Unguentum Plumbi Acetatis.—4; Paraffin Ointment, white, 96.

[1]Liquor Plumbi Subacetatis Fortis. (Goulard's Extract). A clear colourless liquid with a sweet astringent taste and an alkaline reaction.

[1]Liquor Plumbi Subacetatis Dilutis. (Goulard's Lotion or Water.)—1.25; Alcohol, 1.25; Water to 100.

[2]Glycerinum Plumbi Subacetatis.—15; Lead Oxide, 10.5; Glycerin, 50; Water, 36: boiled together and partly evaporated.

[3]Unguentum Glycerini Plumbi Subacetatis.—16.5; Paraffin Ointment, white, 82.5.

[2]Plumbi Carbonas. A hydroxycarbonate, $2(PbCO_3), Pb(OH)_2$. A soft heavy white powder, insoluble in water.

[3]Unguentum Plumbi Carbonatis.—10; Paraffin Ointment, white, 90.

[3]Plumbi Iodidum. Lead Iodide, $PbI_2$. A heavy bright yellow, tasteless and odourless powder. Soluble 1 in 200 of boiling water.

[3]Emplastrum Plumbi Iodidi.—10; Lead Plaster, 80; Resin, 2.

[3]Unguentum Plumbi Iodidi.—10; Paraffin Ointment, yellow, 90.

[1]Plumbi Oxidum. Lead Oxide (Litharge). Heavy pale yellowish-red scales. Insoluble in water.

[1]Emplastrum Plumbi. Lead Plaster.—25; Olive Oil, 50; Water, 25; boiled gently for several hours. An oleate of lead is formed.

**Podophylli Rhizoma. Podophyllum Rhizome.** The dried rhizome and roots of Podophyllum peltatum. Dark reddish-brown, smooth or slightly wrinkled, nearly cylindrical pieces, several inches in length and about 1/3 inch in diameter, with enlargements about 2 inches apart, which bear on their upper surface the scar of an ascending stem and on the lower surface numerous roots. Odour, characteristic, taste slightly acrid and bitter.

[1]Podophylli Resina. Dose, ¼-1 gr.; 15-60 mgms. The resinous precipitate formed by pouring a partially evaporated alcoholic extract into acidified water. An amorphous yellow or orange-brown powder, with a bitter taste. Insoluble in water, soluble in alcohol, and in ammonia solution.

[1]Tinctura Podophylli. Dose, 5-15 min.; 0.3-1 c.c. —3.65; Alcohol to 100: by maceration.

**POTASSIUM. POTASSIUM.**
The incompatibles of the salts of potassium do not depend upon the

potassium, but upon the other radicals present and may hence be best found by looking up the incompatibilities of those radicles.

[1]Potassa Caustica. Potassium Hydroxide. (Hydrate, Caustic Potash.) Hard, white pencils or cakes, very deliquescent. Soluble, 2 in 1 of water, 1 in 2 of alcohol.

[1]Liquor Potassæ. Dose, 10-30 min.; 0 6-2 c.c., freely diluted. —6.19; Water, 100. A colourless odourless alkaline liquid.

[3]Potassa Sulphurata. Sulphurated Potash. (Liver of Sulphur.)

A mixture of the Salts of Sulphur, but chiefly Sulphides.

[1]Potassii Acetas. Dose, 10-60 grs.; 6-40 dgms.
Either in white foliaceous satiny masses or in granular particles, very deliquescent, alkaline in reaction. Soluble 1 in ½ of water, 1 in 2 of alcohol.

[1]Potassii Bicarbonas. Dose, 5-30 grs.; 3-20 dgms.
$KHCO_3$. Colourless crystals, with a feebly alkaline, saline taste. Soluble, 1 in 4 of water, insoluble in alcohol. 20 grs. neutralise 14 grs. of citric or 15 grs. of tartaric acid.

[3]Potassii Bichromas. Dose, 1/10-1/5 gr.; 6-12 mgms.
$K_2CrO_3$, $CrO_2$. Large orange-red transparent crystals. Soluble 1 in 10 of water.

[1]Potassii Bromidum. Dose, 5-30 grs.; 1/3-2 gms.
$KBr$. Colourless, crystals, odourless, with a pungent saline taste. Soluble, 1 in 2 of water, 1 in 200 of alcohol.

[1]Potasssii Carbonas. Dose, 5-20 grs.; 3-12 dgms.
$K_2CO_3$. A white crystalline powder, alkaline and caustic to the taste, very deliquescent. Soluble 1 in 1 of water, insoluble in alcohol.

[1]Potassii Chloras. Dose, 5-15 grs.; 3-10 dgms.
$KClO_3$. Colourless crystals with a cool saline taste. Soluble 1 in 16 of cold, 1 in 3 of hot water, 1 in 1,700 of alcohol.

[2]Trochiscus Potassii Chloratis.—3 grs. with the Rose Basis.

[1]Potassii Citras. Dose, 10-40 grs.; ½-2½ gms.
$C_3H_4.OH.(COOK)_3$. A white powder of a feebly acid saline taste, deliquescent. Soluble 1 in 0.6 of water, 1 in 2 of glycerin, 1 in 9 of alcohol 60%.

[1]Potassii Iodidum. Dose, 5-20 grs.; 3-12 dgms.
$KI$. Colourless crystals with a pungent saline and subsequently bitter taste. Soluble, 1 in less than 1 of water, 1 in 10 of alcohol, 1 in 3 of glycerin.

[3]Linimentum Potassii Iodidi cum Sapone.—10.125; Curd Soap, 13.5; Glycerin, 6.75; Oil of Lemon, 0.844; Water, 67.5.

[3]Unguentum Potassii Iodidi.—10; Potassium Carbonate, 0.6; Water, 9.4; Benzoated Lard, 80

[1]Potassii Nitras.    (Nitre, Saltpetre.)    Dose, 5-20 grs.; 3-12 dgms.
KNO₃. White crystalline masses, colourless with a cool, saline taste. Soluble 1 in 4 of cold, 1 in ½ of hot water, sparingly soluble in alcohol.

[1]Potassii Permanganas.    Dose, 1-3 grs.; ½-2 dgms.
Dark purple slender iridescent crystals, with sweet astringent taste. Soluble 1 in 20 of cold, 1 in 3 of hot water.

[3]Liquor Potassii Permanganatis.    Dose, 2-4 fl. dr.; 8-16 c.c. —1; Water, 100.

[3]Potassii Sulphas.    Dose, 10-40 grs.; ½-2½ gms.
Colourless, hard crystals. Soluble 1 in 10 of cold, 1 in 4 of hot water, insoluble in alcohol.

[3]Potassii Tartras.    Dose, 30-240 grs.; 2-16 gms.
Normal potassium tartrate, $(CHOH)_2.(COOK)_2, H_2O$. Small, colourless crystals. Soluble 1 in 0.6 of water, insoluble in alcohol.

[1]Potassii Tartras Acidus.    (Bitartrate, Purified Cream of Tartar.)    Dose, 20-60 grs.; 1½-4 gms.
A white gritty powder, or fragments of cakes crystallised on one surface, with an acid taste. Soluble 1 in 200 of water, insoluble in alcohol.

PRUNI VIRGINIANÆ CORTEX. The bark of Prunus serotina. In curved pieces or irregular fragments about 1/12 inch thick. Taste astringent aromatic, and bitter; when macerated an odour of bitter almonds, due to the formation of hydrocyanic acid from the amygdalin under the action of the ferment, emulsin.

[1]Syrupus Pruni Virginianæ.    Dose, 30-60 min.; 2-4 c.c.
A macerate and percolate of the bark in water to which sugar and glycerin are added.

[3]Tinctura Pruni Virginianæ.    Dose, 30-60 min.; 2-4 c.c. —20; Alcohol, 62.5; Water, 37.5: by maceration.

*Prunum. Prunes.* The dried ripe fruit of Prunus domestica. Contained in Confectio Sennæ.

*Pterocarpi Lignum. Red Sanders Wood.* (Red Sandal Wood.) The heart wood of Pterocarpus santalinus. Usually imported in logs. In chips it varies in colour from blood-red with lighter zones to a very dark brown; a slight astringent taste and a faint odour when warmed. The colouring matter is soluble in alcohol, and to a very slight extent in water. Used in the Compound Tincture of Lavender.

PULVERES. (The following, with a dose of 1-4 grs. Elaterini Comp.; dose 1-5 grs. Hydrargyrum cum Creta; with dose, 3-6 grs. Antimonialis; dose, 2-10 grs. Opii Comp.; dose 5-15 grs. Ipecacuanhæ Comp.; dose,

5-20 grs. Kino Comp.; dose, 10-20 grs. Scammonii Comp.; dose, 10-40 grs. Cretæ Aromaticus cum Opio, Cinnamomi Comp.; dose, 10,60 grs. Cretæ Aromaticus; dose, 20-40 grs. Catechu Comp.; dose, 20-60 grs. Jalapæ Comp., Rhei Comp., Tragacanthæ Comp.; dose, 60-120 grs. Glycyrrhizæ Comp.; dose, (60-120 grs.). Amygdalæ Comp.; Sodæ Tartaratæ Effervescens, q.v.)

*Pyrethri Radix.* The dried root of Anacyclus Pyrethrum. Pieces 2-4 inches in length, ½ inch or more in diameter, nearly cylindrical or tapering towards each end: outer surface brown and wrinkled; odour characteristic, taste pungent and exciting a copious flow of saliva.

Tinctura Pyrethri.—20; Alcohol 70% 100: by percolation.

**Pyroxylinum. Pyroxylin.** (Di-nitro-cellulose). Soluble readily in equal parts of ether and alcohol.

Collodium. [1]Collodion.—2; Ether, 72; Alcohol, 24.

[1]Collodium Flexile.—96; Canada Turpentine, 4; Castor Oil,2·

[1]Collodium Vesicans. Blistering Collodion.—2.5; Blistering liquid, 100.

**Quassiæ Lignum.** The wood of the trunk and branches of Picræna excelsa. Imported in logs. Retailed in chips of a yellowish-white colour, very light and with a persistently bitter taste. As this bitter contains no tannin it may be prescribed with iron.

[1]Infusum Quassiæ. Dose, ½-1 fl. oz.; 15-30 c.c.-1; Water, 100.

[3]Liquor Quassiæ Concentratus. Dose, ½-1 fl. dr.; 2-4 c.c. —10; Alcohol, 100; an alcoholic extract.

[1]Tinctura Quassiæ. Dose, 30-60 min.; 2-4 c.c. —10; Alcohol 45%, 100: by maceration.

*Quillaiæ Cortex. Quillaia Bark.* (Panama Bark). The inner bark of Quillaja Saponaria. In large flat pieces, 1/6 inch thick and 2 feet or more long, the outer surface is brownish; the inner, smooth and white or yellowish-white: the taste is astringent and acrid, the powder irritating to the nostrils.

Tinctura Quillaiæ. Dose, 30-60 min.; 2-4 c.c. —5; Alcohol, 100; by percolation.

Quinina (see Cinchona p. 51).

Resina. Resin. The residue left after the distillation of the oil of turpentine from the crude oleo-resin of various species of Pinus. Translucent, of a light amber colour, compact, brittle, pulverisable. Soluble in alcohol, ether, benzol, and carbon disulphide.

[1]Emplastrum Resinæ.—10 Lead Plaster, 80; Hard Soap, 5.
[3]Unguentum Resinæ.—28; Yellow Beeswax, 28; Olive Oil by weight, 28; Lard, 21.

**Rhei Radix. Rhubarb Root.** Dose, if repeated 3-10 grs.; 2-6 dgms.: for a single administration 15-30 grs.; 1-2 gms.
The erect rhizome or so-called root of Rheum palmatum, R. officinale and probably other species, deprived of more or less of their cortex and dried. In hard, compact, more or less irregular pieces, smooth, usually reddish-brown or greyish, marked with darker lines and with small scattered star-like marks; odour characteristic somewhat aromatic; taste bitter and feebly astringent. The important pharmacological constituents are the tannoid bodies, of which there are several and the purgative principle, rheopurgarin.

[2]Extractum Rhei. Dose, 2-8 grs.; 1-5 dgms.
An alcoholic percolate evaporated.

[3]Infusum Rhei. Dose, ½-1 fl. oz.; 15-30 c.c.
—5; boiling Water, 100.

[3]Liquor Rhei Concentratus. Dose, 30-60 min.; 2-4 c.c.
—50; Alcohol, 20% to 100 :by percolation.

[1]Pilula Rhei Composita. Dose, 4-8 grs.; 2½-5 dgms. (In 1 or 2 pills).
—27; Socotrine Aloes, 20.25; Myrrh, 13.5; Hard Soap, 13.5; Oil of Peppermint, 1.69; Syrup of Glucose, q.s.

[1]Pulvis Rhei Compositus. (Gregory's Powder.) Dose, 20-60 grs.; 1.2-4 gms.
—22; Light Magnesia, 66; Ginger, 11.

[1]Syrupus Rhei. Dose, ½-2 fl. dr.; 2-8 c.c.
—5; Coriander Fruit, 5; Sugar, 60; Alcohol, 20; Water, 60.

[1]Tinctura Rhei Composita. Dose, if repeated, ½-1 fl. dr.; 2-4 c.c.: for a single administration, 2-4 fl. drs.; 8-15 c.c.
—10; Cardamon Seeds, 1.23; Coriander Fruit, 1.25; Glycerin, 10; Alcohol 60% to 100.

*Rhœados Petala. Red-poppy Petals.* The fresh petals of Papaver Rhœas. The fresh petals are bright scarlet, with a characteristic somewhat unpleasant odour, and a slightly bitter taste.

Syrupus Rhœados. Dose, 30-60 min.; 2-4 c.c.
—22.75; Sugar, 63; Alcohol, 4.375; Water to 100.

*Rosæ Gallicæ Petala.* The dried and fresh unexpanded petals of Rosa gallica. The fresh petals are purplish-red; occur in little cone-shaped masses; odour fragrant; taste somewhat bitter, feebly acid and astringent. Contains a small quantity of tannic acid.

Confectio Rosæ Gallicæ. Fresh Petals, 25; Sugar, 75.

Infusum Rosæ Acidum. Dose, ½-1 fl. oz.; 15-30 c.c. —2.5; Diluted Sulphuric Acid, 1.25; Water, boiling, 100.

Syrupus Rosæ. Dose, 30-60 min.; 2-4 c.c. —4; Sugar, 60; boiling Water, 40.

*Oleum Rosæ.* (*Otto of Rose.*) The oil distilled from the fresh flowers of Rosa damascena. A pale yellow crystalline semi-solid with the strong fragrant odour of rose and a sweet taste.

*Aqua Rosæ.* Prepared by the distillation from the flowers of Rosa damascena.

[3]Unguentum Aquæ Rosæ.—42; Beeswax, white, 9; Spermaceti, 9; Almond Oil by weight, 54; Oil of Rose, 0.1.

Rosmarinus (See Oleum Rosmarani p. 77).

*Saccharum Lactis. Milk Sugar.* (Lactose.) In crystalline masses, with a sweet taste.

**Saccharum Purificatum. Refined Sugar.** (Succrose.) Referred to through this book as sugar. Obtained from the Sugar-cane, sweet, crystals. Soluble 2 in 1 of water.

[1]Syrupus. Syrup.—66.6; boiling Water, to produce 100 by weight.

[3]Syrupus Aromaticus. Dose, 30-60 min.; 2-4 c.c. —50: Cinnamon Water, 25; Tincture of Orange, 25.

[2]Syrupus Glucosi. Syrup of Glucose.—66.6; Liquid commercial glucose, 33.3.

**Salicinum. Salicin.** Dose, 5-20 grs.; 3-12 dgms. A crystalline glucoside obtained from the bark of various species of Salix and Populus. Colourless, shining crystals with a bitter taste; soluble 1 in 28 of water, 1 in 60 of alcohol.

**Salol. Salol.** Dose, 5-15 grs.; 3-10 dgms. Phenyl salicylate. $C_6H_4OH.COOC_6H_5$. Colourless crystals, with a faint aromatic odour and a slight taste. Almost quite insoluble in water, soluble 1 in 12 of alcohol, 1 in 10 of liquid paraffin, and in fixed and volatile oils, very slightly soluble in glycerin.

**Acidum Salicylicum.** Dose, 5-20 grs.; 3-12 dgms. Oxybenzoic acid, $C_6H_4.OH.COOH$. Colourless crystals, taste at first sweetish then acid and leaving a burning sensation in the mouth. Soluble 1 in about 500 of cold, 1 in 15 of hot water, 1 in 3 of alcohol, 1 in 200 of glycerin. Incompatibles, carbonates ($CO_2$ freed), lead acetate, silver nitrate, ferric salts (violet colour produced), potassium iodide, and chlorate, spirits of nitrous ether, quinine sulphate; damp powders or liquids are formed when triturated with lead acetate, sodium phosphate and antipyrine.

[3]Unguentum Acidi Salicylici.—2; Paraffin Ointment, white, **98.** Sodii Salicylas (see p. 95).

*Sambuci Flores. Elder Flowers.* The flowers of Sambucus nigra. Small, flowers, with a slightly bitter taste, and a sweet faint and not altogether agreeable odour.

Aqua Sambuci.—100; Water, 500; distil over 100.

Santal (see Oleum Santali p. 77).

**Santoninum. Santonin.** Dose, 2-5 grs.; 1-3 dgms.
A bitter principle prepared from the flowers of Artemesia maritima, var. Stechmanniana. Colourless, flat crystals, bitter. Scarcely soluble in cold water, sparingly in boiling ,1 in 40 of alcohol.

[2]Trochiscus Santonini. 1 gr. with the Simple Basis.

SAPO ANIMALIS. CURD SOAP. A soap made with sodium hydroxide and a purified animal fat consisting principally of stearin; contains about 30% of water. White, or almost so, dry, nearly inodorous; becomes horny and pulverisable when dried.

SAPO DURUS. HARD SOAP. Soap made with sodium hydroxide and olive oil; contains about 30% of water. Sodium oleate, greyish-white, dry inodorous; becomes horny and pulverisable when dry.

[2]Pilula Saponis Composita. Dose, 2-4 grs.; 12-25 dgms. (In 1 or 2 pills).
—60; Opium, 20; Syrup of Glucose, 20. About 2/5 gr. opium in each pill.

[2]Emplastrum Saponis.—15; Lead Plaster, 90; Resin, 2.5.

SAPO MOLLIS. SOFT SOAP. Made with potassium hydrate and olive oil. Yellowish-white or green, almost inodourous, and of an unctuous consistence.

[1]Linimentum Saponis.—9; Camphor, 4.5; Oil of Rosemary, 1.69; Alcohol, 72; Water, 18.

SARSÆ RADIX. SARSAPARILLA. The dried root of Smilax ornata. Very tough, long flexible roots, brownish in colour, odourless, taste slight and bitter.

[2]Extractum Sarsæ Liquidum. Dose, 2-4 fl. dr.; 8-15 c.c.
—100; Glycerin, 10; Alcohol, 20% sufficient to produce 100: by repeated percolation.

[3]Liquor Sarsæ Compositus Concentratus. Dose 2-8 fl. dr.; 8-30 c.c.
—100 ; Sassafras Root, 10 ; Guaiacum Wood, 10; Dried Liquorice Root, 10;

Mezereon Bark, 5; Alcohol, 22.5; Water, q.s.; by repeated infusion, extraction and evaporation.

*Sassafras Radix.* The dried root of Sassafras officinale. In large branched pieces more or less covered with a rough brownish bark: odour agreeable; peculiar, aromatic, astringent taste.

SCAMMONIÆ RADIX. SCAMMONY ROOT. The dried root of Convolvulus Scammonia. Greyish tapering or nearly cylindrical roots, often contorted and longitudinally furrowed: enlarged at the crown and bears the remains of slender aerial stems; odour characteristic, taste, at first somewhat sweet, afterwards acrid.

SCAMMONIÆ RESINA. SCAMMONY RESIN. Dose, 3-8 grs.; 2-5 dgms. Prepared by extracting the root with alcohol and precipitating the resin with water. Brownish translucent pieces, brittle, with a sweet odour.

[2]Pilula Scammonii Composita. Dose, 4-8 grs.; 2½-5 dgms. (In 1 or 2 pills.)
—32; Jalap, Resin, 32; Curd Soap, 32; Tincture of Ginger, 96.

[3]Pulvis Scammonii Compositus. Dose, 10-20 grs.; 6-12 dgms.
—50; Jalap, 37.5; Ginger, 12.5.

SCAMMONIUM. SCAMMONY. Dose, 5-10 grs.; 3-6 dgms. A gum resin obtained by the incision of the living root of Convolvulus Scammonia, often called Virgin Scammony. In flattened cakes or irregular pieces, dark grey to black externally and covered with a greyish powder, very brittle and the resin within is more or less porous, glossy, and nearly black: odour characteristic; taste, acrid.

**Scilla. Squill.** Dose, 1-3 grs.; 6-18 cgms.
The bulb of Urginea Scilla, divested of its dry membranous outer scales and dried. The slices of the inner scales usually present the form of curved strips frequently tapering towards the ends, yellowish or pinkish, somewhat translucent; inodorous, disagreeably bitter.

[2]Acetum Scillæ. Dose, 10-30 min.; 0.6-2 c.c.
—12.5; Diluted Acetic Acid, 100: by maceration.

[1]Syrupus Scillæ. Dose, 30-60 min.; 2-4 c.c.
—34.5; Sugar, 65.5: product should weigh 100.

[3]Oxymel Scillæ. Dose, 30-60 min.; 2-4 c.c.
—6.75; Acetic Acid, 6.75; Water, 21.6; Clarified Honey to 100.

[3]Pilula Scillæ Composita. Dose, 4-8 grs.; 2½-5 dgms. (In 1 or 2 pills.)
—25; Ginger, 20; Ammoniacum, 20; Hard Soap, 20; Syrup of Glucose, q.s.

[3]Pilula Ipecacuanhæ cum Scilla (see Ipecacuanha).
[2]Tinctura Scillæ.   Dose, 5-15 min.; 0.3-0.6 c.c.
—20; Alcohol, 60%, 100: by maceration.


SCOPARII CACUMINA.   BROOM TOPS.   The fresh and dried tops of Cytisus scoparius.
[1]Infusum Scoparii.   Dose, 1-2 fl. oz.; 30-60 c.c.
—10; boiling Water, 100.
[3]Succus Scoparii.   Dose, 1-2 fl. dr.; 4-8 c.c.
75 of juice of fresh tops; Alcohol, 25.


SENEGÆ RADIX.   SENEGA ROOT.   The dried root of Polygala Senega. Greyish or yellowish slender roots, 2-4 inches long, enlarged at the top into a crown bearing the basis of numerous slender aerial stems: odour distinctive; taste at first sweet, subsequently acrid.
[1]Infusum Senegæ.   Dose, ½-1 fl. oz.; 15-30 c.c.
—5; boiling Water, 100.
[3]Liquor Senegæ Concentratus.   Dose, 30-60 min.; 2-4 c.c.
The extract of 50 in 100 of diluted alcohol.
[3]Tinctura Senegæ.   Dose, 30-60 min.; 2-4 c.c.
—20; Alcohol 60%, 100: by percolation.


**Senna Alexandrina.   Alexandrine Senna.**   The dried leaflets of Cassia acutifolia. Pale, greyish-green, brittle leaflets, ¾-1¼ inch in length, unequal at the base: odour, faint, peculiar; taste mucilaginous and unpleasant.

**Senna Indica.   East Indian Senna.**   (Tinnivelly Senna.) The dried leaflets of Cassia angustifolia. Pale yellowish-green leaflets unequal at the base; odour and taste similar to the above. The following preparations may be made from either of these two drugs.

[3]Confectio Sennæ.   Dose, 60-120 grs.; 4-8 gms.
—9.33; Coriander Fruit, 4; Figs, 3.16; Tamarinds, 12; Cassia Pulp, 12; Prunes, 8; Extract of Liquorice, 1.33; Sugar, 40; Water sufficient to produce 100 by weight.
[1]Infusum Sennæ.   Dose, ½-1 fl. oz.; 15-30 c.c.: as a draught, 2 fl. oz.; 60 c.c.
—10; Ginger, 0.63; boiling Water, 100.
[2]Mistura Sennæ Composita.   (Black Draught.)   Dose, 1-2 fl. oz.; 30-60 c.c.
Magnesium Sulphate, 25; Liquid Extract of Liquorice, 5; Compound Tincture of Cardamons, 10; Aromatic Spirit of Ammonia, 5; Infusum of Senna to 100.

[3]Liquor Sennæ Concentratus. Dose, 30-60 min.; 2-4 c.c. —100; Tincture of Ginger, 12.5 ; Alcohol, 10 ; Water, to 100: by repeated percolation.

[1]Syrupus Sennæ. Dose, ½-2 fl. dr.; 2-8 c.c. —40; Oil of Coriander, 0.02; Alcohol, 0.08; Sugar, 50; Alcohol, 20%, 70; Water, q.s.: by repeated maceration, evaporation, and solution.

[2]Tinctura Sennæ Composita. Dose, if repeated, 30-60 min.; 2-4 c.c.: for a single administration 2-4 fl. dr.: 8-16 c.c. —20; Raisins, 10; Caraway Fruit, 2.5; Coriander Fruit, 2.5; Alcohol 45%, 100: by maceration.

*Serpentariæ Rhizoma. Serpentary Rhizome.* The dried rhizome and roots of Aristolochia Serpentaria or of A. reticulata. The rhizome of A. Serpentaria is tortuous and slender about 1 inch long and 1/8 in diameter, bears on its upper surface remains of aerial stems and on the lower numerous wiry roots. A. reticulata is similar but the rhizome is larger. Odour camphoraceous, taste strong, bitter, and aromatic.

Infusum Serpentariæ. Dose, ½-1 fl. oz.; 15-30 c.c. —5; boiling Water, 100.

Liquor Serpentariæ Concentratus. Dose, ½-2 fl. dr.; 2-8 c.c. —50; Alcohol, 20%, 100: by percolation.

Tinctura Serpentariæ. Dose, 30-60 min.; 2-4 c.c. —20; Alcohol, 70% 100: by percolation.

*Sevum Præparatum. Prepared Suet.* The internal fat of the abdomen of the sheep, Ovis aries, prepared by melting and straining.

*Sinapis. Mustard.* The dried ripe seeds of Brassica nigra and B. alba powdered and mixed. A greenish yellow powder, with a bitter pungent taste, and a pungent odour when moist.

[1]*Charta Sinapis. Mustard Paper.* The bruised seeds are percolated with benzol to remove the fixed oil. The residue is dried, powdered, mixed with the solution of India rubber and spread upon cartridge paper and dried.

*Sinapis Albæ Semina. White Mustard Seed.* The dried ripe seeds of Brassica alba. Very small, pale yellow seeds, in taste less pungent than the black mustrad. ·

*Sinapis Nigræ Semina. Black Mustard Seed.* The dried ripe seeds of Brassica nigra. Very small, dark brown, spherical seeds: taste bitter and pungent: odourless until ground with water, when the sinigrin is broken up by the ferment myrosin, and the pungent volatile oil, allylisothiocyanate is produced.

[2]Oleum Sinapis Volatile. Distilled from the seeds after maceration with water. Colourless or pale yellow oil, with an intensely penetrating odour and an acrid taste; almost immediately vesicates the skin.

[3]Linimentum Sinapis.—4; Camphor, 6; Castor Oil, 14; Alcohol, 86.

## SODIUM. SODIUM. The metal of commerce.
The incompatibles of the salts of sodium do not depend upon the sodium but upon the other radicles present and may be found by looking up these.

[1]Soda Tartarata. Sodium Potassium Tartarate. (Rochelle Salt.) Dose, 120-240 grs.; 8-16 gms.
$(CHOH)_2COONa.COOK$, $4H_2O$. Crystals, colourless, with a saline taste. Soluble 1 in 1½ of water, insoluble in alcohol.

[1]Pulvis Sodæ Tartaratæ Effervescens. (Seidlitz Powder.) Dose, the two powders mixed in water. (1)-120; Sodium Bicarbonate, 40 grs. (in blue paper). (2) Tartaric Acid, dried, 38grs. (in white paper).

[2]Sodii Arsenas. Sodium Arsenate. Dose, 1/40-1/10 gr.; 1.5-6 mgms.
An anhydrous disodium hydrogen arsenate, $Na_2HAsO_4$. A white powder. Soluble 1 in 6 of water, (solution has an alkaline reaction). Slightly soluble in alcohol.

[3]Liquor Sodii Arsenatis. Dose, 2-8 min.; 0.1-0.8 c.c.
—1; Water, 100. An alkaline solution.

[2]Sodii Benzoas. Sodium Benzoate. Dose, 5-30 grs.; ¼-2 gms. A white, crystalline or amorphous powder, inodorous, or with a faint odour of benzoin, and an unpleasant sweetish saline taste. Soluble 1 in 2 of water, 1 in 24 of alcohol.

[1]Sodii Bicarbonas. (Baking Soda). Dose, 5-30 grs.; ¼-2 gms. $NaHCO_3$. In powder or small, white crystals, with a saline taste. Soluble 1 in 11 of water, insoluble in alcohol.

[3]Trochiscus Sodii Bicarbonatis. 3 grs. with the Rose Basis.

[1]Sodii Bromidum. Sodium Bromide. Dose, 5-30 grs.; ¼-2 gms.
$NaBr$. Small, white crystals somewhat deliquescent, inodorous, with a saline taste. Soluble 1 in 1.2 of water, 1 in 16 of alcohol.

[1]Sodii Carbonas. (Washing Soda). Dose, 5-30 grs.; ½-2 gms. $Na_2CO_3,10H_2O$. Transparent, colourless crystals, efflorescent, with a harsh taste, and a strongly alkaline reaction. Soluble 1 in 2 of water. 10 parts neutralise 4.9 parts of citric acid or 5.75 parts of tartaric acid.

[1]Sodii Carbonas Exsiccatus. Dose, 3-10 grs.; 2-6 dgms.
Sodium carbonate freed from its water of crystallization by heat. A white powder. 53 gr. equivalent to 14.3 gr. of crystallized salt.

[1]Sodii Chloridum. Purified common salt.

[2]Sodii Citro-tartras Effervescens. Dose, 60-120 grs.; 4-8 gms.
Sodium Bicarbonate, 51; Tartaric Acid, 27; Citric Acid, 18; Sugar, 15.

[2]Sodii Hypophosphis. Dose, 3-10 grs.; 2-6 dgms.
$NaPH_2O_2$. A white deliquescent granular salt, with a bitter nauseous taste. Soluble, 1 in 1 of water, 1 in 20 of alcohol.

[1]Sodii Iodidum. Dose, 5-20 grs.; 3-12 dgms. NaI. A dry white crystalline powder, with a saline and somewhat bitter taste. Soluble 1 in 0.55 of water, 1 in 3 of Alcohol, 1 in 1 of glycerin.

[1]Sodii Nitris. Dose, 1-2 grs.; 6-30 cgms.
A white deliquescent crystalline powder. Soluble 1 in 1.2 of water, 1 in 50 of alcohol.

[1]Sodii Phosphas. Dose, if repeated, 30-120 grs. 2-8 gms.; for a single administration, ¼-½ oz.; 7-14 gms.
Di-sodium-hydrogen-phosphate, $Na_2HPO_4$, $12H_2O$. Transparent, colourless crystals, efflorescent, with an alkaline reaction and a saline taste. Soluble 1 in 6 of water, insoluble in alcohol.

[2]Sodii Phosphas Effervescens. Dose, if repeated 60-120 grs.; 4-8 gms.: for a single administration, ¼-½ oz.; 7-14 gms.
—50; Sodium Bicarbonate, 50; Tartaric Acid, 27; Citric Acid, 18. The sodium phosphate should be dessicated before using.

[1]Sodii Salicylas. Dose, 10-30 grs.; ½-2 gms.
In small, colourless scales or crystals, with a pearly lustre; taste sweetish, but unpleasantly saline, odourless. Soluble 1 in 1 of water, 1 in 5 of alcohol.

[1]Sodii Sulphas. Dose, if repeated, 30-120 grs.; 2-8 gms.: for a single administration, ¼-½ oz.; 7-14 gms.
Transparent crystalline, efflorescent salt, with a bitter saline taste. Soluble 1 in 2.8 of water at 15°C.; 1 in 0.3 at 33° C.; insoluble in alcohol.

[2]Sodii Sulphas Effervescens. Dose, if repeated, 60-120 grs.; 4-8 gms.: for a single administration, ¼-½ oz.; 7-14 gms.
—50; Sodium Bicarbonate, 50; Tartaric Acid, 27; Citric Acid, 18. The sodium sulphate should be dessicated before using.

[2]Sodii Sulphis. Dose, 5-20 grs.; 3-12 dgms. $Na_2SO_3$, $7H_2O$.
Colourless, transparent, efflorescent crystals, inodorous, with a saline and sulphurous taste. Soluble 3 in 4 of water, insoluble in alcohol.

[2]Sodii Sulphocarbolas. Sodium Sulphocarbolate. Dose, 5-15 grs.; 3-10 dgms.

Sodium phenol-para-sulphonate, $C_6H_4OH.SO_2ONa,2H_2O$. Colourless, transparent crystals, with a saline and somewhat bitter taste. Soluble 1 in 6 of water.

Liquor Sodii Ethylatis. Contains 18% of sodium ethylate, $NaC_2H_5O$. A colourless syrupy liquid made by cautiously dissolving sodium in absolute alcohol. Incompatible with water.

SPIRITUS. (With a dose of 5-20 min. Anisi, Cajuputi, Camphoræ, Cinnamomi, Lavandulæ. Menthæ Piperitæ, Myristicæ; dose of 5-20 or 30-40 min. Chloroformi; dose, 20-60 min. Juniperi; dose, 20-40 or 60-90 min. Ætheris, Ætheris Comp., Ætheris Nitrosi, Ammoniæ Aromaticus, Ammoniæ Fetidus; dose, 1-2 fl. dr. Armoraciæ Comp.; without stated dose, Rosmarini (5-30 min.), Rectificatus, Vini Gallici.)

**Spiritus Aetheris Nitrosi. Spirit of Nitrous Ether.** (Sweet Spirit of Nitre). Dose, if repeated, 20-40 min.; 1½-2½ c.c.: for a single administration, 60-90 min.; 4-6 c.c.

An alcoholic solution containing ethyl nitrite, aldehyde and other substances. An inflammable limpid liquid, of a faint yellow colour, and with a peculiar, penetrating apple-like odour, and a characteristic taste. Often slightly acid in reaction due to free nitrous acid.

Incompatibles, alkali hydrates; if acid hypophosphites, sulphites, chlorates, iodides, ammonium bromide, mercurous salts, permanganates, antipyrine, acetanilid, salicylates, tannic acid, thymol, morphine guaiacum, acetates.

Spiritus Rectificatus. (See Alcohol p. 31).

*Staphisagriæ Semina. Stavesacre* Seeds. The dried ripe seeds of Delphinium Staphisagria. Irregularly triangular or quadrangular, brownish seeds, with no marked odour and a nauseous, bitter acrid taste.

Unguentum Staphisagriæ.—20; Yellow Beeswax, 10; Benzoated Lard, 85.

STRAMMONII FOLIA. The dried leaves of Datura Strammomium. Dark greyish-green, wrinkled leaves, 4-6 inches long, with a characteristic odour, and an unpleasant bitter taste.

[1]Tinctura Strammonii. Dose, 5-15 min.; 0.3-1 c.c. —20; Alcohol 45% to 100.

STRAMMONII SEMINA. The dried ripe seeds of Datura Strammonium Dark brown or nearly black, flattened, reniform seeds about 1/6 of an inch long; surface is pitted; no marked odour, but a slightly bitter taste.

[1]Extractum Strammonii. Dose, ¼-1 gr.: 15-60 mgms. An evaporated alcoholic percolate.

**Strophanthi Semina. Strophanthus Seeds.** The dried ripe seeds of Strophanthus Kombe. Oval acuminate, flattened seeds about 3/5 inch long, covered with silky hairs, greenish-fawn in colour: odour characteristic, taste very bitter. The active principle is the glucoside strophanthin.

[2]Extractum Strophanthi. Dose, ¼-1 gr.; 15-60 mgms.
An alcoholic percolate evaporated and to which milk-sugar has been added.

[1]Tinctura Strophanthi. Dose, 5-15 min.; 0.3-1 c.c.
—2.5; Alcohol 70% 100: by percolation.

Strychnina (see Nux Vomica p. 75).

*Styrax Præparatus. Prepared Storax.* A balsam obtained from the trunk of Liquidambar orientalis purified. A semi-transparent, brownish-yellow, semi-liquid balsam, with a strong agreeable odour and balsamic taste.

*Succi.* (With dose, 5-15 min. Belladonnæ; dose, 30-60 min. Hyoscyami; 1-2 fl. dr. Conii, Scoparii, Taraxaci; (1-2 fl. oz.), Limonis.)

SULPHONAL. SULPHONAL. Dose, 10-30 grs.; ½-2 gms.
Dimethyl-methane-diethylsulphone, $(CH_3)_2C(SO_2C_2H_5)_2$. Colourless, inodorous, nearly tasteless crystals. Soluble, 1 in 450 of cold, 1 in 15 of boiling water, 1 in 50 of alcohol.

**Sulphur. Sulphur.**
Incompatibles, triturated with strong oxidising agents, such as potassium permanganate or chlorate, an explosion is apt to occur.

[1]Sulphur Præcipitatum. Precipitated Sulphur. (Milk of Sulphur.) Dose, 20-60 grs.; 1.2-4 gms.
A greyish-yellow soft powder, free from grittiness and from the smell of hydrogen sulphide.

[2]Trochiscus Sulphuris.—32.4; Acid Potassium Tartrate, 6.48; Sugar, 51.84; Gum Acacia, 6.48; Tincture of Orange, 5.9; Mucilage of Acacia, 5.9; 100 Trochisci, the quantities of the solids in grammes, those of the liquids in cubic centimetres; 5 gr. in each lozenge.

[1]Sulphur Sublimatum. Sublimed Sulphur. (Flowers of Sulphur.) Dose, 20-60 grs.; 1.2-4 gms.
A bright greenish yellow slightly gritty powder, without taste or odour.

[3]Confectio Sulphuris. Dose, 60-120 grs.; 4-8 gms.
—40; Acid Potassium Tartrate, 10; Tragacanth, 0.4; Syrup, 20; Tincture of Orange, 5; Glycerin sufficient to produce 100 by weight.

[2]Unguentum Sulphuris.—10; Benzoated Lard, 90.

[3]Sulphuris Iodidum. Sulphur Iodide. A greyish-black, solid substance, with the odour of iodine.

[3]Unguentum Sulphuris Iodidi.—4; Glycerin, 4; Benzoated Lard, 92.

*Sumbul Radix. Sumbul Root.* The dried transverse slices of the root of Ferula Sumbul. Pieces varying much in size, covered on the outer surface with a brown, papery, wrinkled cork, often beset with short hairs; internally, spongy, dry, fibrous and yellowish-brown; odour strong and musk-like; taste bitter and aromatic.

Tinctura Sumbul. 30-60 min.; 2-4 c.c.
—10; Alcohol 70%, 100.

*Suppositoria* (see Acidum Carbolicum, Acidum Tannicum, Belladonna, Glycerinum, Iodoformum, Morphina, Plumbum.)

SYRUPI. (Without stated dose Syrupus and Syrupus Glucosi; with a dose of ½-1 fl. dr. Aromaticus, Aurantii, Aurantii Floris, Calcii Lactophosphatis, Ferri Iodidi, Ferri Phosphatis, Ferri Phosphatis cum Quinina et Strychnina, Hemidesmi, Limonis, Pruni Virginianæ, Rhœados, Rosæ, Tolutanus, Zingiberis; with dose, ½-2 fl. dr. Cascaræ Aromaticus, Chloral, Codeinæ, Rhei, Sennæ.)

TABELLÆ TRINITRINI. TRINITRIN TABLETS. Dose, 1-2 tablets. Tablets of chocolate weighing 5 grs. each containing 1/100 gr. of Nitroglycerine.

*Tamarindus. Tamarinds.* The fruits of Tamarindus indica, freed from the brittle outer part and preserved in sugar.

*Taraxaci Radix. Taraxacum Root.* The fresh and dried roots of Taraxacum officinale. The fresh roots are frequently more than a foot in length, break readily and from the broken surface a milk exudes. The dried root is shrivelled and wrinkled, dark brown in colour. Inodorous, taste bitter.

[2]Extractum Taraxaci. Dose, 5-15 grs.; 3-10 dgms.
The juice of the fresh root dried to a soft consistence.

[1]Extractum Taraxaci Liquidum. Dose, ½-2 fl. dr., 2-8 c.c.
An extract in diluted alcohol made by maceration and partial evaporation.

[3]Succus Taraxaci. Dose, 1-2 fl. dr.; 4-8 c.c.
Fresh juice, 75; Alcohol, 25.

*Terebenum. Terebene.* Dose, 5-15 min.; 0.3-1 c.c.
A mixture of hydrocarbons. A colourless liquid, with an agreeable odour and an aromatic taste.

*Terebinthina Canadensis. Canada Turpentine.* (Canada Balsam.) The oleo-resin obtained from Abies balsamea. Pale yellow and faintly greenish, transparent oleo-resin of the consistence of thin honey.

*Thus Americanum. Frankincense.* The concrete oleo-resin scraped off the trunks of Pinus palustris and P. Tæda.

THYMOL. THYMOL. Dose, ½-2 grs.; 3-12 cgms.
A crystalline substance obtained from the volatile oils of Thymus vulgaris, Monarda punctata, and Carum copticum. Large, colourless crystals with the odour of thyme and a pungent aromatic taste. Almost insoluble in water, freely soluble in alcohol.

Incompatibles, spirit of nitrous ether; gives a soft or liquid mass when triturated with, acetanelid, antipyrine, camphor, phenol, chloral, menthol, quinine sulphate, resin, salol.

**Thyroideum Siccum. Dry Thyroid.** Dose, 3-10 grs.; 2-6 dgms.
A powder prepared from the fresh and healthy thyroid of the sheep.

LIQUOR THYROIDEI. THYROID SOLUTION. Dose, 5-15 min.; 0.3-1 c.c. A watery extract of the fresh and healthy thyroid glands of the sheep, containing 0.5% of phenol as a preservative. A pinkish, turbid liquid, entirely free from any odour of putrescence.

TINCTURÆ (see Aconitum, Aloe, Arnica, Asafetida, Aurantium, Belladonna, Benzoinum, Buchu, Calumba, Camphora, Cannabis Indica, Cantharis, Capsicum, Cardamomum, Cascarilla, Catechu, Chirata, Chloroformum, Cimicifuga, Cinchona, Cinnamomum, Coccus, Colchicum, Conium, Crocus, Cubeba, Digitalis, Ergota, Ferrum, Gelsemium, Gentiana, Guaiacum, Hamamelis, Hydrastis, Hyoscyamus, Iodum, Jaborandi Jalapa, Kino, Krameria, Lavandula, Limon, Lobelia (etherial), Lupulus, Myrrha, Nux Vomica, Opium, Podophyllum, Prunus Virginiana, Pyrethrum, Quassia, Quillaia, Quinina, Rheum, Scilla, Senega, Senna, Serpentaria, Strammonium, Strophanthus, Sumbul, Balsamum Tolutanum, Valeriana, Zingiber). 30-60 min. is the dose of all tinctures except the following: (1) having a dose of 5-15 min. Belladonna, Cannabis Indica, Capsicum, Chloroform and Morphine, Coccus, Colchicum, Digitalis, Iron Perchloride, Lobelia, Nux Vomica, Opium, Podophyllum, Squills, Strammonium, Strophanthus; (2) Having a dose of 2-5 min. Iodum, Aconite, and Cantharides, the two latter may for a single administration be given in 5-15 min. doses.

TRAGACANTHA. A gummy exudation obtained from the stem of Astragulus gummifer, and other species, white or pale yellowish flakes, flattened, thin, irregular and marked on the surface by concentric rings; somewhat translucent, horny, inodorous and almost tasteless. Sparingly soluble in water, but swells into a mass with it.

[1]Glycerinum Tragacanthæ.—**20**; Glycerin, **60**; Water, **20**: by trituration.

¹Mucilago Tragacanthæ.—1.38; Alcohol, 2.25; suspend the former in the alcohol; and add Water to 100.
¹Pulvis Tragacanthæ Compositus. Dose, 20-60 grs.; 1.2-4 c.c. —16.5; Gum Acacia, 16.5; Starch, 16.5; Sugar, 49.5.

Trinitrin (see Tabellæ Trinitrini p. 98. Liquor p. 72)

*Trochisci* (see Acidum Benzoicum, Acidum Carbolicum, Acidum Tannicum, Bismuthum, Catechu, Eucalyptus, Ferrum, Guaiacum. Ipecacuanha, Krameria, Morphina, Potassii Chloras, Santoninum, Sodii Bicarbonas, Sulphur.)
The Trochisci, Lozenges, are made with three bases of which the chief constituents are Gum Acacia, Mucilage of Acacia and Sugar. The Simple Basis consists of these ingredients alone, the Fruit Basis contains black current paste as flavouring: the Rose Basis contains Rose Water, and the Tolu Basis, Tincture of Tolu.

*Unguenti* (see Acidum Boricum, Acidum Carbolicum, Acidum Salicylicum, Aconitina, Rosa, Atropina, Belladonna, Cantharis, Capsicum, Cetaceum, Chrysarobinum, Cocaina, Conium, Creosotum, Eucalyptus, Galla, Plumbum, Hamamelis, Hydrargyrum, Iodum, Iodoformum, Paraffinum, Pix, Plumbum, Potassium, Resina, Staphisagria, Sulphur. Veratrina, Zincum.)

UVÆ URSI FOLIA. BEARBERRY LEAVES. The dried leaves of Arctostaphylos Uva-ursi. Yellowish-green coriaceous leaves, about 3/4 inch long, the upper surface glabrous, shining and reticulate, with depressed veinlets, no definite odour but a very astringent taste. The active principle, is a glucoside, arbutin.
Infusum Uvæ Ursi. Dose, ½-1 fl. oz.: 15-30 c.c. —5; boiling Water, 100.

**Valerianae Rhizoma. Valerian Rhizome.** (Root.) The dried erect Rhizome and roots of Valeriana officinale. A short erect rhizome often sliced, yellowish-brown externally, with numerous slender roots: odour strong characteristic and disagreeable; taste, unpleasant camphoraceous, and slightly bitter.

¹Tinctura Valerianæ Ammoniata. Dose, 30-60 min.; 2-4 c.c. —20: Oil of Nutmeg, 0.31; Oil of Lemon, 0.21; Solution of Ammonia, 10; Alcohol 60%, 90.

*Veratrina. Veratrine.* An alkaloid prepared from the dried ripe seeds of Schœnocaulon officinale. Pale grey, amorphous, odourless but powerfully irritant to the nose, strongly and persistently bitter and intensely acrid.
Unguentum Veratrinæ.—2; Oleic Acid, 8; Lard, 90.

VINI. (With a dose of 10-30 min. or 2-4 fl. dr. Antimoniale. Dose 10-30 min. or 4-6 fl. dr. Ipecacuanhæ; dose, 10-30 min. Colchici: dose, 1-4 fl. dr. Ferri Citratis; ½-1 fl. oz. Quininæ; without stated dose. Aurantii, Xericum.)

### Zincum. Zinc.

Incompatibles of soluble salts in solution, hydrates, carbonates, phosphates arsenates, borax, tannic acid, albumin.

[1]Zinci Acetas. Zinc acetate. Dose, 1-2 grs.; 6-12 cgms. Translucent, colourless, crystals with a pearly lustre, and a sharp unpleasant taste. Soluble 1 in 2½ of water.

[2]Zinci Carbonas (Calamine). A hydroxycarbonate, $ZnCO_3$, $(ZnH_2O_2)_2$, $H_2O$. A white tasteless, odourless powder. Insoluble in water.

[1]Zinc Chloridum. $ZnCl_2$. Colourless, opaque rods or tablets very deliquescent and caustic. Soluble 10 in 4 of water and in alcohol.

[3]Liquor Zinei Chloridi. Colourless liquid with a sweetish astringent taste.

[1]Zinci Oxidum. Dose, 3-10 grs.; 2-6 dgms. A soft, nearly white, tasteless and inodourous powder.

[1]Unguentum Zinei.—15; Benzoated Lard, 85.

[1]Zinci Sulphas. White Vitriol. Dose, as a tonic, 1-3 grs.; ½-2 dgms: as an emetic, 10-30 grs.; 6-20 dgms. $ZnSO_4$, $7H_2O$. Colourless, transparent crystals, with a strong metallic styptic taste. Soluble, 1 in less than 1 of water.

[2]Zinci Sulphocarbolas. Zinc Sulphocarbolate. Zinc phenolparasulphonate. Colourless, transparent, efflorescent crystals. Soluble 1 in 2 of water, 1 in 2½ of alcohol.

[3]Zinci Valerianas. Dose, 1-3 grs.; ½-2 dgms. Pearly white tabular crystals with a disagreeable odour and a metallic taste, very slightly soluble in cold water, soluble in hot water and alcohol.

[2]Unguentum Zinci Oleatis. A precipitate due to the mixture of solutions of Zinc Sulphate and Hard Soap, washed and dried and then mixed with Soft Paraffin.

ZINGIBER. GINGER. The dried and scraped rhizome of Zingiber officinale. Flat, irregular branched pieces, varying in length: odour agreeable aromatic, taste hot and pungent.

[1]Syrupus Zingiberis. Dose, 30-60 min.; 2-4 c.c. A strong tincture, 5; Syrup to 100.

[1]Tinctura Zingiberis. Dose, 30-60 min.; 2-4 c.c. —10; Alcohol, 100: by percolation.

# CHAPTER VI.

## NON-OFFICIAL MATERIA MEDICA.

The author has selected a few of the newer remedies as illustrations of the non-official materia medica. In his selection he has been guided by two considerations, the degree of popularity that the drugs enjoy and the likelihood of their proving to be permanently useful. Examples of some of the non-official methods of galenical preparation such as the elixers have also been included as it was considered that they would be of use to the student. The arrangement of these drugs is the same as in the proceeding chapter and the source for them has largely been the British Pharmaceutical Codex.

ACETOMORPHINÆ HYDROCHLORIDUM. ACETOMORPHINE HYDROCHLORIDE. Dose, 1/40-1/6 gr.; 2-10 mgms.
An alkaloid obtained from morphine by the substitution of two H. groups by acetyl groups, diacetyl-morphine-hydrochloride. White crystals, soluble in water, 1 in 2, in alcohol 1 in 9. (Trade-name, Heroin.)

ACIDUM CRESYLICUM. CRESYLIC ACID. (Cresol.) Dose, 1-3 min.; 0.05-0.2 c.c.
A mixture of ortho-, meta-, and para-cresols. Crude cresol is a yellowish liquid, darkening on keeping, with a characteristic odour resembling phenol. Soluble 1 in 80 of water, readily in alcohol, ether, chloroform, glycerin, olive oil. Pure ortho-cresol is a colourless, deliquescent, crystalline solid.

ACIDUM ACETYL-SALICYLICUM. ACETYL-SALICYLIC ACID. Dose, 8-15 grs.; ½-1 gm. (Salacetic Acid). $C_6H_4(COOH)OCOCH_3$. A white crystalline powder, or colourless crystals. Soluble 1 in 400 of water. Soluble 1 in 5 of alcohol. Both aqueous and alcoholic solutions do not keep on standing. (Trade names, Acetysal, Aletodin, Aspirin, Saletin, etc.)

**ADRENINA.** (Epinephrine. Nephridine.) Dioxy-phenyl-ethanol-methylamine, $C_6H_3(OH)_2CHOHCH_2NHCH_3$. The active principle of the suprarenal gland (nephridium, adrenal, epirenal); it may also be produced synthetically. A drab or buff coloured, minutely crystalline powder. Decomposes in the presence of water, and especially in the weak alkalies. Soluble 1 in 5000 of ether or alcohol, readily soluble in water acidulated with hydrochloric acid. (Trade names for the active principle of the suprarenal, Adrenalin, Adrenin, Epirenan, Haemostasin, Hemisine, Suprarenin, etc.).

[1]Ammonium Bromidum. Ammonium Bromide. Dose, 5-30 grs.; 3-20 dgms.
Colourless crystals or a white crystalline powder. Soluble 2 in 3 of water 1 in 15 of alcohol.

**Argyrol.** (Silver Vitellin.) A compound of silver and a vegetable protein containing about 30% of silver. Readily soluble in water (the solution decomposes on keeping).

Boroglycerinum. Boric Acid, 47; Glycerin by weight, 64. Differs from the official Glycerin of Boric Acid in the proportions and the method of making. It contains glyeryl borate, which readily breaks down in the presence of water. A white viscid opaque liquid of a honey-like consistence. Readily soluble in water and alcohol.

[1]Calcii Lactas. Calcium Lactate. Dose, 10-60 grs.; ½-4 gms.
White granular masses, or powder, or in crystals, odourless and with scarcely any taste. Soluble 1 in 15 of water, scarcely soluble in alcohol.

CARBASUS ABSORBENS. Absorbent Cotton. Open-wove cotton gauze or mulls prepared from cotton freed from its natural oil.
[2]Carbasus Acidi Borici. Boric Acid Gauze.—100; Saturated Solution of Boric Acid tinted with aniline red a sufficient quantity. The gauze is immersed in the boiling solution and subsequently dried. It should contain 40-50% of boric acid.
[2]Carbasus Iodoformi. A gauze impregnated with 10% of iodoform.

[2]Cataplasma Kaolini. Kaolin Poultice. Kaolin, 52.7; Boric Acid, 4.5; Thymol, 0.05; Methyl Salicylate, by weight, 0.2; Oil of Peppermint, by weight, 0.05; Glycerin, by weight, 42.5. Heat the kaolin to 100°C. and maintain at that temperature for one hour, occasionally stirring, add the boric acid, mix intimately, incorporate the glycerin, finally add the thymol dissolved in the methyl-salicylate and the oil of peppermint. The mixture should be kept warm for four hours with occasional stirring, and preserved in air-tight vessels.
[1]Cataplasma Lini. Linseed Poultice. Linseed, crushed, 28; boiling water, 100. Add the linseed gradually to the boiling water stirring constantly.
[1]Cataplasma Sinapis. Mustard Poultice. Linseed, crushed, 28; Mustard, powdered, 2; Water to produce 100. Add the linseed to about 70 of water, then add the mustard, previously rubbed to a paste with water.

[2]Ceratum Galeni. Galen's Cerate. (Cold Cream.) Soft Paraffin, white, 12; White Beeswax, 12; Almond Oil, 50; Borax, 1; Oil of Rose, 0.10; Rose Water, 25.

**DIOXYDIAMINO-ARSENOBENZOL HYDROCHLORIDUM**
(Ehrlich-Hata). Dose 0.2-0.7 gm., $HCl.NH_2OH.C_6H_3: As.C_6H_3.OH.NH_2$-HCl is a yellowish powder which dissolves slowly in water giving a clear acid solution. Neutralized with normal solution of caustic soda a precipitate occurs which again dissolves on the addition of more alkali when the sodium salt of dioxydiamino-arsenobenzol $NaO.NH_2C_6H_3As: C_6H_3.NH_2ONa$ has been formed. This salt is used for both intravenous and intramuscular injection. It is not stable and must be prepared fresh as required. For intravenous infusion the slight excess of alkali formed, is neutralized with 1% acetic acid and diluted with normal saline. (Trade name Salvarsan; "606").

[2]Elixer Aromaticum. Aromatic Elixer. Dose, ½-2 fl. dr.; 2-8 c.c. Compound Spirit of Orange, 2.5; Syrup, 37.5; Purified Talc, 3; Alcohol, a sufficient quantity; Water to 100. Filtered.

[2]Elixer Aurantii. Elixer of Orange, 15-60 min.; 1-4 c.c. Oil of Bitter Orange, 0.3; Alcohol, 30; Syrup, 35; Cinnamon Water to 100.

[2]Elixer Bismuthii. Elixer of Bismuth. Dose, ½-1 fl. dr.; 2-4 c.c. Bismuth and Ammonium Citrate, 3.3; Distilled Water, hot, 6; Solution of Ammonia a quantity sufficient (to keep the mixture clear); Aromatic Elixer to 100.

[1]Emulsio Olei Morrhuæ. Emulsion of Cod-liver Oil. Dose, ¼-1 fl. oz.; 8-30 c.c. Cod-liver Oil, 50; Gum Acacia, 12.5; Syrup, 6.25; Oil of Bitter Almonds, 0.1; Water to 100.

Enema Aloes. Aloes, 0.75; Potassium Carbonate, 0.25; Glycerin 10; Mucilage of Starch to 100. 10 fluid ounces are used.

Enema Opii. Tincture Opium 3. Mucilage of Starch to 100. 2 ounces are used.

Enema Terebinthinæ. Oil of Turpentine, 2; Mucilage of Starch to 100. 16 fluid ounces are used.

*Ergotoxina. Ergotoxine.* Dose, 1/12-1/6 gr.; 5-10 mgms. An alkaloid, an active principle of ergot. A light, white, amorphous powder. Practically insoluble in water. Its salts are, however, soluble.

**Ethylis Chloridum. Ethyl Chloride.** $C_2H_5Cl$. A colourless, mobile liquid, with a sweetish burning taste, and an agreeable odour. Boiling-point 12.5°C.

ETHYLMORPHINÆ HYDROCHLORIDUM. ETHYLMORPHINE HYDROCHLORIDE. Dose, 1/10-½ gr. 6-30 mgms. A white minutely crystalline powder, odourless with a bitter taste. Soluble 1 in 7 of water, 1 in 5 of alcohol. (Trade name, Dionin.)

**Eucalyptol. Eucalyptol.** Dose, 1-5 min.; 0.06-0.3 c.c.
A purified substance prepared from the oil of eucalyptus and other sources. A colourless liquid, with a characteristic aromatic camphoraceous odour and a spicy pungent taste, leaving a cool sensation in the mouth. Miscible in all proportions with alcohol, but not with water.

²Gargarisma Acidi Tannici. Tannic Acid Gargle. Glycerin of Tannic Acid. 10; Water to 100.
²Gargarisma Aluminis. Alum Gargle. Alum, 2; Acid Infusion of Roses to 100.
²Gargarisma Boracis. Borax, 4; Water to 100.
²Gargarisma Potassii Chloratis. Potassium Chlorate, 2; Diluted Hydrochloric Acid, 1; Water to 100.

²Gossipium Acidi Borici. Boric Acid Wool. Cotton Wool immersed in a saturated solution of boric acid tinged with aniline red and removed and dried. Contains 40-50% of boric acid.

**Guaiacol. Guaiacol.** Dose, 1-5 min.; 0.06-03 c.c.
$C_6H_4OCH_3OH$. Obtained either synthetically or by the fractional distillation of wood creosote. An oily colourless liquid, with a penetrating, smoky odour and a caustic taste. Soluble 1 in 80 of water, miscible with alcohol ether and oils.

²Guaiacolis Carbonas. Guaiacol Carbonate. Dose, 5-15 grs.; 3-10 dgms. May be gradually increased to 30 grs.; 2 gms. The carbonic ester of guaiacol. A white crystalline powder almost without taste or odour. Soluble 1 in 70 of alcohol, insoluble in water.

LIQUOR FORMALDEHYDI. A solution of formaldehyde obtained by dissolving in water formic aldehyde. A colourless transparent liquid with a pungent odour and a caustic taste. Miscible with water and alcohol. Should contain 38-39% of aldehyde.

**Hexamethylenamina. Hexamethylene-tetramine.** (Hexamethylenamine, Formamine.) Dose, 5-15 grs.; 3-10 dgms.
$(CH_2)_6N_4$. A white crystalline powder, odourless; in solution has an alkaline reaction. Soluble 1 in 1½ of water, 1 in 8 of alcohol (Urotropin).

METHYL SALICYLICUM. METHYL SALICYLIC ESTER (Oil of Wintergreen). Dose 5—7 min. 0.3—0.5 c.c.
A light yellow oily fluid, soluble in fats, alcohol, and ether.

**Methylthioninæ Hydrochloricum.** (Methylene Blue.) Dose, 1-5 grs.; ½-3 dgms.
A dull dark-green crystalline powder. Soluble 1 in 50 of water, less soluble in alcohol.

[2]Mucilago Amyli. Mucilage of Starch, Starch 1.5; Water, to 100.

**Parahydroxyphenylethylamine.** (Dose 20 mgms.) soluble in water, probably an important active constituent of preparation of ergot. (Trade name Tyramine.)

PELLETIERINÆ TANNAS. PELLETIERINE TANNATE. Dose, 5-8 grs.. 3-5 dgms.
The tannate of an alkaloid obtained from the root-bark of pomegranate, Punica Granatum. A light yellow amorphous powder, greyish-white, odourless, with an astringent taste and an acid reaction. Soluble 1 in 700 of water, 1 in 80 of alcohol.

**Phenolphthaleinum. Phenolphthalein.** Dose, 1-8 grs.; ½-5 dgms.
A white or almost white amorphous or crystalline powder, odourless. Soluble 1 in 800 of water, 1 in 10 of alcohol. Trade names are many e.g. Laxans, Laxoin, Laxophen, Probilin, Purgen, etc.

STRONTII BROMIDUM. STRONTIUM BROMIDE. Dose, 5-30 grs.; ¼-2 gms.
$SrBr_2, 6H_2O$. Colourless transparent odourless crystals, deliquescent, and with a strong bitter saline metallic taste. Soluble 2 in 1 of water, 1 in 3 of alcohol.

**SERUM ANTIDIPHTHERICUM. ANTIDIPHTHERIC SERUM.** (Diphtheria Antitoxine.) Dose, as a prophylactic, 500 units; as a curative agent, 2,000-4,000 units or more.
A unit is the quantity of antitoxine necessary to prevent the death of a guinea-pig weighing 250 grammes when injected with 100 lethal doses of diphtheria toxin. The serum is the blood-serum of horses immunised by the injection at stated intervals with diphtheria toxin in amounts at first sublethal but finally many times the lethal dose. The blood is drawn from the horse under the most careful aseptic precautions, is allowed to clot, the serum removed and set aside for several weeks during which time a precipitate forms which is filtered off. The serum is now put up in suitable containers, usually with the addition of some antiseptic, such as phenol or cresol. Its antitoxic power is tested previous to its being placed in the containers and into each of these latter is put a definite number of units. The number of units and the date of the preparation of the serum must be placed on a label upon each container. The serum decreases in activity with age, losing 10-30% per annum. A dried serum is also prepared.

**SERUM ANTITETANICUM. ANTITETANIC SERUM.** (Tetanus Antitoxin.) This is similarly prepared to the antidiphtheric serum, the horses being injected with tetanus toxin.

SERUM ANTISTREPTOCOCCICUM. ANTISTREPTICOCCIC SERUM. Dose 30 c.c. daily subcutaneously.
The serum obtained from the blood of a horse immunised by repeated injections of at first killed cultures of Streoptococcis pyogenes or S. erysipelatus of many different stems, and later of virulent living culture.

SERUM ANTIMENINGOCOCCICUM. A serum prepared in an analogous manner by injections of the Diplococcus intracellularis, and used in the treatment of Cerebro-spinal Meningitis.

**Theobromina. Theobromine.** Dose, 5-10 grs.; 3-6 dgms.
A dimethylxanthin obtained from the seeds of Theobroma Cacao. A white crystalline powder odourless with a bitter taste. Soluble 1 in 1,700 of water, 1 in 1,400 of alcohol. A compound with sodium and sodium salicylate, Theobromine Sodio-salicylate, has a sweetish taste and is more soluble, 1 in 1 of water, and is hence used in place of the alkaloid alone. (It is sold under the trade name of Diuretin.)

TERPINI HYDRAS. TERPIN HYDRATE. (Terpene Hydrate.) Dose, 3-10 grs.; 3-6 dgms.
An alcoholic hydrate prepared from oil of turpentine. Colourless, glistening crystals or a crystalline powder, with a slight aromatic odour and a bitter taste. Soluble 1 in 280 of water, 1 in 14 of alcohol.

**Tuberculinum. Tuberculin.** (Old.) Three months old glycerine—veal broth cultures of the Bacillus tuberculosis are concentrated over a water-bath and filtered through a porcelain filter to remove the bacilli. For treatment New Tuberculin or other similar preparations are more largely used.

**TUBERCULINUM NOVUM. NEW TUBERCULIN.** (Tuberculin R.) The Bacillus tuberculosis is grown on glycerine-serum and the resulting cultures scraped off and heated to 60° C. to kill the bacteria, dried in vacuo and triturated. The resulting mass is emulsified and centrifuged, the upper layer rejected and the lower again dried, triturated and again emulsified and centrifuged, the upper layer is set aside and the residue subjected to the same process. The resulting layers are preserved with glycerine and standardised to contain 10 milligrammes of solids, in 1 c.c. It may be used thus as a liquid or may be dried. The dose is calculated as 1/6,000-1/1,000 mgms. of solid substance.

# CHAPTER VII.

## NOTES ON PRESCRIBING.

As each patient must be considered as an individual and given separate care and thought, it is impossible in such a book as this, which does not treat of therapeutics, to do more than indicate some of the points which must be understood and borne in mind when about to write a prescription.

Drugs are administered by the physician either for a local effect, that is, produced in the immediate neighborhood of the point of application, or for their remote effect, that is, produced generally on various organs throughout the body, or on some special organ remote from the point of application for which they have a special affinity.

It is consequently essential for the prescriber to decide in the first place what organs he wishes to influence and whether they are such as he can reach by local application or not. Then in the second place, he must decide what drugs he desires to use to affect these organs.

**For Application to the Skin.** Very few drugs can be absorbed into the system through the skin with such rapidity that they accumulate within the body sufficiently to be used by the physician for their remote pharmacological action. The outstanding example of those that do so is mercury. Most other drugs are excreted by the kidney and bowels so much more readily than they are absorbed from the skin that no remote action can be expected.

The rapidity with which drugs may penetrate the skin can be greatly increased by dissolving them in some solvent which penetrates more readily than they do themselves. Such solvents are used as the bases of those ointments from which we wish an action after absorption. Alcohol, to a slight extent, olive-oil, wool-fat, and lard, are all absorbed by the skin. Olive-oil is perhaps the best base to aid absorption. Wool-fat is very nearly as good, while lard is not very efficient. The paraffins, resins, and soaps, are scarcely absorbed at all, and solution in them rather delays than furthers the absorption of substances made into ointments or plasters with them.

A local effect, but one produced in the skin after absorption, is naturally more easily obtained. Oil of Mustard developed in a mustard poultice or plaster,—Cantharidin from its Collodion, Liquor, Emplastrum, Unguentum, or Tincture,—Croton Oil, Turpentine, Ammonia and Chloroform, readily pass through the skin and by reflex stimulation set up a more or less marked local inflammation and possibly a remote reflex effect. The student will note that the bases of Unguentum Cantharidis are more soluble in the skin than those of the Emplastrum and consequently contain a lower percentage of Cantharis. Atropine, morphine, cocaine, camphor, potassium iodide and salicylates all pass through the skin in quantities sufficient to produce local deep effects if they are combined with suitable bases such as the oils, wool-fat, and alcohol, and especially if they are aided by heat or rubbing or both. Such bodies as ammonia, chloroform, camphor, and turpentine, which by a local action after absorption lead to a dilatation of the vessels of the skin, aid in the more rapid absorption of such other bodies as can penetrate the skin, and this forms one reason for their inclusion in ointments, liniments, plasters. If the student, with these ideas in mind, will examine the pharmacopœial liniments and ointments he will gain a good idea of how he may perscribe for local and reflex effects by way of the skin.

Lead Plaster is largely used for mechanical and supporting purposes, but none of the other pharmacopœial plasters save those of Cantharis and possibly Belladonna are frequently employed.

An antiseptic in the form of boracic acid 5%, salicylic acid ¼%, or benzoic acid 1%, is frequently added to dusting powders consisting of talc, starch, or zinc oxide, or to ointments containing some such inert protective powder, (e.g. Unguentum Zinci Oxidi).

In many cases a purely superficial effect on the skin is desired. Antiseptics must often be applied to the skin in order to kill pediculi or disease germs which are lying upon it. For this purpose phenol in ½-2% watery solution, mercuric chloride 1-500-2000, boracic acid in a saturated solution, are amongst the common drugs used. If the skin be abraded, very much weaker solutions of phenol or mercury must be applied, otherwise sufficient antiseptic would be absorbed to produce a remote effect, or they may be replaced by less readily absorbed antiseptics such as iodoform, ichthyol, thymol, or resorsin. In any such case, one of the above drugs may be applied in the form of an ointment whose base is not readily absorbed by the skin.

A very considerable insight into the character of preparations intended for external use may be gained by the examination of the examples of liniments and ointments and of the preparations of mercury. If the constituents and bases of the following preparations of mercury are examined it may be seen that the Emplastrum Hydrargyri, Emplastrum Ammoniaci cum Hydrargyro, Unguentum Hydrargyri Oxidi Flavi, Unguentum Hydrargyri Oxidi Rubri, Unguentum Hydrargyri Ammoniati, and the Lotio Hydrargyri Flava vel Nigra would be relatively slowly absorbed; the Unguentum Hydrargyri Compositum, Unguentum Hydrargyri Iodidi Rubri, Unguentum Hydrargyri Oleatis, Unguentum Hydrargyri Subchloridi, more readily; while the Unguentum Hydrargyri Nitratis and Linimentum Hydrargyri will be still more readily taken up. The first group of these would therefore be more generally used for purely external and local purposes where they would act as antiseptics. The last group might readily be used for remote and specific effects.

From the inlets of the body covered with mucous membranes, drugstuffs are much more readily absorbed than from the skin. Drugs applied to them can readily be made to exert a remote action.

**For application to the rectum.** Enemata are sometimes used to produce remote action. This method is only resorted to when the drugs to be given are either too unpalatable to be taken per as or when they would irritate the mouth or stomach of the patient. Drugs given per rectum are usually administered in, roughly, twice the dose in which they would be taken by the mouth. If the drug-stuffs contained in an enema are to be absorbed it is necessary that they should not be irritant to the mucous membrane nor should its bulk be so large as to mechanically set up movements. Hence watery or weak alcoholic mixtures 1-2 ounces in bulk are usually used. Medicaments which are very readily absorbed may be given in the form of a suppository. Three of the pharmacopœial suppositories contain drugs which exert remote actions. These are, Suppositoria Belladonnæ, Morphinæ, Plumbi.

By means of an enema antiseptics may be applied to the surface of the rectum and a part of the colon. If the antiseptic is readily absorbed or powerful in its action on mucous membranes, e.g. Acidum Carbolicum, only a small quantity of it may be prescribed in a large bulk of water. If relatively insoluble, e.g. Argyrol, or Boracic Acid, a much stronger solution may be used. Enemata containing bitters and astringents, —Quassia, Tannic Acid, Kino, are sometimes applied to drive out pin-worms and to act as mild astringents.

An enema is frequently employed to soften hardened faeces or to bring about defæcation. For these purposes either some bland fluid, e.g. Olive Oil, Normal Saline, or Mucilage of Starch is used in large quantities (2-3 pints) as these mechanically dilate the bowel and set up reflex movements. In order to produce defæcation alone a much smaller quantity of fluid may be used if it contains some drug which acts as a local irritant to the sensory nerve endings, e.g. Turpentine, Mustard, Aloes.

Nourishment is often given to those whose stomach is deranged, by means of an enema. In this case a bulk of more than three or four ounces can rarely be given. The food must be fluid, nonirritant, and highly nutritous. Eggs, milk, oil, alcohol, or mixtures of these with water, are common ingredients of this type of enema. In order to aid in their retention the fluid should be warmed to the temperature of the body, should be given slowly, and the patient kept in a prone position.

Some of the pharmacopœial suppositories are intended to produce local antiseptic or astringent actions in the bowel, e.g. those of Phenol, Iodoform, Tannic Acid, and Lead; that of Glycerin to aid in defæcation.

**For application to the Vagina.** Drugs are only given by the Vagina for their local effect. Suppositories for this purpose are usually known as Pessaries,—fluid washes as Douches. The most frequently employed vaginal antiseptics are Mercuric Perchloride 1-2000-5000, Silver Nitrate 2-1000-500, Argyrol, and Cresylic Acid, Potassium Permanganate 1-1000-5000. Vaginal Tampons impregnated with active drugs in glycerin are sometines employed.

**For application to the Urethra and Bladder.** Here also a purely local effect is the only one ever sought. Astringents and antiseptics are administered to the urethra and bladder in the form of douches whose solvent is some bland fluid, usually water or Oil. Suppositories for the urethra are known as Bougies. The antiseptics mentioned above for the vagina are very commonly employed here also.

As the bulk of fluid necessary for a douche is very often a large one it is a common expedient of prescribers to order for their patient either powders or concentrated solutions to which large quantities of water such as can be readily measured in the household, pints or quarts, are to be added as needed.

**For application to the Conjunctiva.** This is the most delicate of all the mucous surfaces and in consequence as Collyria only weak solutions of astringents and antiseptics may be employed and usually the weaker members of these series are chosen,—Sulphate of Zinc 1-250, Silver Nitrate 1-200-500, Argyrol and other colloidal preparations of Silver may be used in stronger solutions up to 5%, Boracic Acid and Borax in 2% solutions. Solutions of these salts are frequently made with Camphor Water as a vehicle.

In order to produce their local effects after absorption, the Mydriatics and Miotics may be prescribed, as the Lamellæ or more usually in solution, Atropine in 1%, Cocaine in 5%, Physostigmine in ½%, Homatropine in 2%, Pilocarpine in ½-1%, Ethylmorphine in 1%.

All solutions especially if they contain inorganic salts should be carefully filtered so as to make certain that they are free from gritty particles of dirt, which would irritate the sensitive mucous surface.

Drugs may also be applied to the Conjunctiva in solution in oil or in the form of an ointment; such bases must be bland, free from fatty acids and from insoluble particles. Liquid and yellow (not white) Soft Paraffin, conform best to these requirements.

**For application to the mouth and Respiratory Passages.** Antiseptic and astringent solutions are frequently given in the form of douches, gargles (Gargarismata), mouth washes (Collutoria). The chief antiseptics used are Boracic Acid, Borax, Potassium Chlorate, all in about 4% solution,—Eucalyptol and Thymol. Of the astringents the Liquor or Tincture of Ferric Chloride 2-3%, Tannic Acid 1-2%, Tincture of Kino 2% Alum 1%, Potassium Permanganate 1-1000, are frequently used preparations. Antiseptics or astringents are often applied by means of a swab and in this case stronger solutions may be used, e.g. Silver Nitrate ½-1-2%.

The Larynx and upper parts of the Nose can also be reached by means of douches. For the nose the solution should be a bland one if a thorough application is to be hoped for. A saline solution of the concentration of normal saline which contains as well mild antiseptics such as Thymol and Eucalyptol or a solution of one of the colloidal silver preparations are very popular. Such solutions may be very well applied by means of an atomizer. Solutions for use in an atomizer are very frequently made with Liquid Paraffin as a solvent.

The trachea, bronchioles, and alveoli, can only be reached by a very fine spray or by volatile substances which can be inhaled. Most of the antiseptics which on account of their volatility could be applied in this way, are either too irritant or so readily absorbed that they would produce un-wished-for remote actions. And though good may be done in some cases by applying the weaker antiseptics it is rarely that they can be brought to the diseased area in sufficient concentration or for sufficient duration of time, to produce any marked effect. The antiseptics that may be used in this way include Benzoin, Thymol, Eucalyptol, Creosote. The Compound Tincture of Benzoin is a favorite preparation. Volatization is usually brought about by pouring a strong alcoholic solution of the antiseptic upon the surface of boiling water and inhaling the fumes which arise. In this way, not only the antiseptic, but also water vapour which serves to allay the feeling of dryness of the inflamed mucous membrane, is inhaled. Volatile substances such as Ether, Chloroform, and Amyl nitrite readily produce a remote action when inhaled.

**Administration by the mouth.** For remote action drugs are most commonly given by the mouth though as pointed out above they may be administered by the rectum, the skin, or the lungs. The use of hypodermic intramuscular and intravenous injections is increasing. When the physician has decided to give the drugs required by the mouth, several pharmaceutical forms may be used; mixtures are still the most commonly employed, though pills are extensively used when the drugs are unpleasant and the dose is small; powders are used only when one wishes to administer larger doses of tasteless or not markedly unpleasant drugs in quantities larger than can be given in pill form; caches and capsules are forms steadily gaining in vogue especially for the administration of drugs having an unpleasant flavor.

MIXTURES. It is good practice for the physicain when he writes a prescription for a mixture to use only such drugs as will dissolve and produce a clear solution. This is a good but by no means an absolute rule and indeed we find in the pharmacopœia very striking deviations from it, e.g. Mistura Ferri. This intentionally contains the incompatibles Iron Sulphate and Potassium Carbonate resulting in the formation of an insoluble precipitate and a murky solution. The Iron Carbonate formed is, however, less irritant to the stomach than the Sulphate. Mixtures containing a precipitate were very commonly prescribed in the past, but today a physician is compelled to pay more attention to the likes and whims of his patients, all of whom have seen and tasted attractive and pleasant patented preparations. In writing a prescription for a mixture the physician should, as a rule, use fluid preparations of the drug selected if such are contained in the Pharmacopœia. The reasons for this are easily seen if one considers the matter from the view-point of the dispenser. Suppose that thirty doses of Strychnine 1/60 of a grain and Arsenious Anhydride 1/40 gr. are to be given. This would force the dispenser, if the solids were prescribed, to weigh out ½ and 30/40 gr. of the two drugs respectively, while if Liquor Arsenicalis 80 min. and Liquor Strychninæ Hydrochloridi 110 min. were prescribed the same amount of each drug would have been given and the dispenser's work made easy and more rapid as Arsenious Anhydride is difficult to get into solution. The Liquors, Spirits, Tinctures, Liquid Extracts, Waters, Syrups, are the important fluid preparations intended for administration in mixtures. When there is a choice of salts the more soluble one should be used.

The physician must take every possible care that his mixtures are as palatable and as pleasant to the eye as possible. The only colorings provided in the Pharmacopœia are Crocus, Cochineal, and Red Sanders Wood, and the Compound Tincture of Cardamons  Neither Crocus nor Cochineal should be used in an acid medium. All of the above produce shades of red. Sweetening may be provided in the form of Simple Syrup or even better as one of the flavored syrups e.g. Aromatic Syrup, Syrup of Orange, of Ginger or of Tolu. Liquorice contains a particularly sweet flavoring principle and is very greatly used. Amongst the Waters and Spirits the attention of the student may be drawn to the usefulness of Chloroform, Cinnamon, Orange, and Peppermint. Rose flavors are more usually used today for lotions and ointments than for mixtures. Acids and Bitters are also in many cases useful flavors. General rules for flavoring are extremely difficult to give and since the taste of each physician and indeed of each of his patient will vary, it is difficult even to give useful hints. For such vegetable drugs as Digitalis, Ergot, Ipecac, Krameria, and the bitter of most alkaloids Syrup of Orange is one of the best flavors, aided perhaps by some water such as that of Cinnamon or Peppermint; for Opium, Ginges forms a good covering: for the salts of Potassium, such flavoring Waters ar Chloroform or Peppermint with Aromatic Syrup may be used; Potassium Iodide and Quinine may be covered with Extract of Liquorice; Sodium Salicylate by Cinnamon Water, and Syrup; Copaiba by Peppermint.

PILLS form perhaps the best method of administering unpleasant drugs whose dose is small. As many people find pills difficult to swallow, they must be made as small as possible, and should never exceed 5 grains in weight and rarely should exceed 3 grains. In consequence of this the student should examine with care the preparations of any drug which he intends to give in pill form and choose the one with the smallest dose. The only exception to this rule should be made when one of the preparations has physical properties which would be of value in forming a pill mass. The prescriber must also bear in mind the fact that unless amongst the drugs that he wishes to prescribe there is one whose physical properties are such as to bind the others together some adhesive substance or excipient must be added by the dispenser and that this will necessarily increase the bulk of these pills to a certain extent. The best excipient for general use is probably Tragacanth either in the form of its Glycerin or Compound Powder, as very small quantities of these are needed. Powdered Hard Soap may be used with powdered vegetable drugs and gum resins, and Curd Soap with essential oils and creosote. As each pharmacist is apt to become familiar with one particular type of excipient and prefer it to all others it is often well to omit the excipient from his prescription, but he must not fail to remember that it must be added and will increase its bulk.

The student will notice that the solid Extracts are in many cases the most compact form in which vegetable drugs can be given and they and the Green Extracts, which are very sticky and form good pill bases, are introduced into the Pharmacopœia as ingredients of pills. It is rarely that aqueous or alcoholic solutions can be incorporated in pills.

POWDERS. A physician, when considering the administration of drugs in powder form, must always carefully consider the flavor of the principal drug and whether if it is unpleasant or even tasteless, its palatability can be increased by adding some flavoring. It is rarely that the taste of a disagreeable powder can be successfully covered. If a disguising flavor is wanted, Sugar, Liquorice and Cinnamon are perhaps amongst the best. The physician must remember that deliquescent salts cannot be given in powder form.

CACHETS AND CAPSULES. These are much used in modern dispensing as they enable the physician to administer disagreeable and bulky powders and also oils in an elegant manner. Roughly speaking one may order up to 10 grains of a drug in capsule form, and up to 20 in a cachet. If the drugs are very heavy these quantities may be readily increased. Fluids, with the exception of oils, should not as a rule be given in capsules.

**Administration of drugs by hypodermic, intramuscular, and intravenous injections.** These methods of administration make certain the complete absorption of the drugs given if they are soluble in the fluids of the body. And in consequence of this much smaller quantities are used than for administration per os. Drugs given by hypodermic injection should be non-irritant to the sensory nerve-endings of the part. They should in consequence be neither acid nor alkaline. When given intramuscularly this point should also be observed, but when given intravenously it becomes of very little importance as the drug does not come in contact with the sensory endings. If drugs are given hypodermically, roughly one half the dose that would be given per os may be used. If given intramuscularly, a slightly smaller quantity is usually given and if given intravenously, owing to the rapidity with which it reaches the point of attack, approximately one-tenth that would be given per os is administered. These rules are by no means absolute as drugs differ so greatly in their physical and pharmacological properties. In the case of all these methods of administration the very greatest care must be taken that the solutions are strictly aseptic and that the patient's skin is carefully sterilized before the administration is undertaken.

# CHAPTER VIII.

## PRESCRIPTION-WRITING.

When the physician has decided upon the drugs which he wishes to administer to a patient, the form, pill, powder mixture, etc., in which he wishes to administer them, and the preparations that are best suited to the form chosen, he has still to write a prescription which will convey his wishes clearly and concisely to the pharmacist. Even if the physician does his own dispensing the writing of a careful prescription is not to be omitted, as it is essential that he have for the purpose of consultation in the future a statement in writing of the treatment adopted, also the writing of a prescription will save many errors in dispensing. The question of the ownership of the prescription is a doubtful one some claiming that it is simply an order by the physician to the pharmacist, who should keep it as a record of the orders given him. On the other hand very many persons hold that the prescription is the property of the patient to whom it is given. The pharmacist can hardly refuse to give the original holder of the prescription a copy thereof, unless he has distinct orders not to do so from the physician. In view of this when the physician writes a prescription which he does not want repeated he should not only mark it "ne repetatur" but should also inform the patient that this prescription is one in which he has no proprietary interest but is only the physician's instructions to the pharmacist. This precaution should always be taken when prescribing morphine in any form. The pharmacist is expected, not only to refuse a copy of any prescription to any person other than the one to whom the physician gave it, but also not to make any further use of it.

The prescription was formerly written entirely in Latin, and even today the great majority of prescriptions are written largely in that language. This custom possesses some distinct advantages. The official Latin names are concise and distinctive so that there is little danger of error. Formerly when Latin was the universal language of science and medicine, it ensured that the prescription could be universally read and understood, this still to a certain extent holds good as most civilised governments have adopted official Latin names in their pharmacopœias, though unfortunately the Latin names adopted differ slightly in different countries.

It is a good rule to write the names of the drugs and the directions to the pharmacist if they be simple and well understood in Latin while the directions to the patient which are to be inscribed by the pharmacist upon the label should be written in English as this ensures that no error will arise in translation. The directions to the dispenser may of course be given in English and indeed it is well to do so if they are at all unusual. Directions to the patient should as a rule be written in English, but there are a few simple directions which have been so much used that the abbreviations of their Latin translation are very commonly employed. The student will often find in older books prescriptions with Latin directions. For these reasons he should make himself familiar with the phrases given in the vocabulary.

*In writing a prescription always write legibly.* Do not endanger the success of your treatment or possibly even the patient's life by careless, illegible hand-writing. Whenever large quantities of any powerful drug are ordered, and especially if they surpass the pharmacopœial dose, the quantities should be not written in numerals but should also be written out in full in words.

The following is a typical prescription:—

*Superscription*
*Inscription*



*Subscription*
*Signature*

The words to the left, inscription, superscription, subscription and signature, are the names applied to those parts of the prescription opposite which they are set. The signature includes the directions to the patient. The other three parts are for the pharmacist. The subscription includes the compounding directions to the pharmacist.

The following is a transcription in unabbreviated Latin of the above prescription with an interlinear translation.

For Arthur H.
Recipe
*Take thou*
Potassii Acetatis..................................unciam unam·
*Of Acetate of Potassium.........................one drachm.*
Liquoris Ammonii Acetatis............drachmas tres cum semisse.
*Of Solution of Acetate of Ammonium........three and a half drachms.*
Spiritus Ætheris Nitrosi......................drachmas duas.
*Of Spirits of Nitrous Ether.................... two drachms.*
Infusi Buchu..........(quantum, sufficiat usque) ad uncias quattuor.
*Of Infusion of Buchu (a quantity sufficient) up to four ounces.........*

Misce. Fiat mistura. Signa:—
*Mix. Let a mixture be made. Label:—*
Drachmam unam ter in die post cibos.
*One drachm three times a day after meals.*

The grammatical form proves on examination not to be a difficult one. The verb "recipe" which is invariably used, governs the accusative. It is clear that the pharmacist is not to take all of his stock of any ingredient but only a part thereof. Hence the nouns expressing the quantity, "unciam" "drachmas" are in the accusative governed by "recipe." The names of the ingredients of which the stated quantities are to be taken are in the partitive genitive. Adjectives must agree with the noun that they modify in gender, number, and case; so "duas" and "tres" agree with "drachmas," "unam" with "unciam," and "quattuor" though indeclinable with "uncias." "Nitrosi" also agrees with Ætheris. Potassii is again in the partitive genitive as are both "Ammonii" and "Acetatis" in the following line.

The last line of the inscription gives slightly more trouble. As usually written the words included within brackets are omitted, yet the clause beginning with "quantum" is the object of the sentence and is governed by the verb "recipe." "Infusi" is again in the partitive genitive. "Quantum" is in the accustive for the reason given; "sufficiat" is the third person singular of the present subjunctive owing to the clause being a subordinate one; "usque" is an adverb meaning "upto" "until"; "ad" a preposition governing the noun "uncias." There is a slightly different form in which this line is occasionally written in which in place of "Infusi" "Infusum" would be written; this is the partitive use of the accusative.

"Misce" like "Recipe" is the second person singular form of the imperative mood of a verb of the second conjugation, while "signa" is the form of the same tense, number, person, and mood of a verb of the first. "Mistura" is the nominative of a noun of the first declension. "Fiat" is the third person singular of the present subjunctive and is an example of the jussive use of that tense as a mild imperative. "Drachmam unam" is the accusative governed by some such verb understood as "capiat" another example of the jussive use of the subjunctive. "Ter" is a numeral adverb. "Die" the ablative of the noun "dies" after the preposition "in." "Cibos" is the accusative plural of "Cibus." Several other similar stereotyped forms are in use in the signature of which the following is one of the more common "Drachma una ter in die sumenda." The translation would be the same. "Drachma" is in the nominative singular and has agreeing with it the gerundive of the transitive verb "sumo." This use of the gerundive signifies duty or necessity and hence an order in a mild form.

The following points in regard to the manner of writing should be noted. The custom has been adopted of writing the numeral expressing the quantity after the abbreviation for the measure. The numeral is written in small Roman numerals except in the case of fractions, or where one wishes to draw special attention to the quantity; in both these cases the Arabic numerals are used. Further the "i's" in the Roman numerals should have a dash above the letter and the dot should be carefully and distinctly written above the dash, so that they may be counted in confirmation of the number of strokes below the dash, should any question arise.

Abbreviations should be used with the greatest care and only such as are certain to be understood. For example such abbreviations, as "chlor" which might mean chloral, chloroform, chloridum, or as "hyd" which might stand for hydrargyrum, hydras, or hydrochloridum, are not permisssible. The usual abbreviations for common words will be found in the vocabulary.

Were the above prescription written in the metric system it would be as follows (in order to fill a standard bottle of 150 c.c. it has been recalculated and now contains 42 doses).

℞                                                              Gm. vel. c.c.

| | | | |
|---|---|---|---|
| Potassii Acet............ | 41.5 gms. | | 41 \| 50 |
| Liq. Ammon. Acet..... | 17.5 c.c. | or | 17 \| 50 |
| Spt. Æth. Nit......... | 10.25 c.c. | | 10 \| 25 |
| Infus. Buchu......... ad | 150.00 c.c. | | 150 \| 00 |

The quantities are as a rule written in Arabic numerals, and the measure if the prescription be not written on paper with a heading as shown on the right follows the numerals as is shown on the left. Fractions are always written as decimals and again paper as printed on the right with a perpendicular line to distinctly mark the decimal point is a great advantage and a great safe-guard. Such prescriptions when read are commonly read in English and not in Latin.

The mathematics involved in prescription writing is not more difficult than is the grammar. Two points must first be decided, (1) For how many days and how many doses a day are you going to give the medicine? Taking the case used above as an illustration, we will suppose that you have decided to give three doses a day for a period of ten days, in all thirty doses. (2) How much of each ingredient do you wish to give at each dose? We will suppose that you intend to give 15 grs. of Potassium Acetate, 7 min. of Solution of Ammonium Acetate, 4 min. of the Spirit of Nitrous Ether, and some of the Infusion of Buchu (the latter is a comparatively inactive flavouring ingredient and may be given in considerable doses). You have already only 11 mins. of fluid; the acetate is very soluble and would readily dissolve in 30 min. therefore there is no need to give a larger dose than 1 fl. dr. The total quantity that you will want is 30 fl. dr. 32 fl. dr. make 4 fl. oz., which is the size of a standard bottle. The prescription will then be written for 32 doses, or of Acetate of Potassium 32 x 15 grs. = 480 grs. one Troy ounce; of Solution of Ammonium Acetate, 32 x 7 = 224 min. or approximately 3½ fl. dr. (210 min.); of Spirits of

Nitrous Ether 4 x 32 = 128 min. approximately 2 fl. dr. Similar calculations may readily be made for any other prescription. It is customary to round off the amounts to make even numbers in drachms or ounces if the drugs be not very potent but if potent this practice should never be followed. The amount prescribed should suffice to fill a standard bottle. The standard bottle sizes are ½, 1, 2, 3, 4, 6, 8, 10, 12, and 16 oz.

PHARMACOPŒIAL NOMENCLATURE.—The principles of the pharmacopœial nomenclature are very simple. As far as possible the Latin names are direct equivalents of the English names. . . . The names of acids may be considered as direct translations of the English names. For example "Hydrochloric Acid, becomes "Acidum Hydrochloricum." "Acidum" is a neuter noun of the second declension with a genitive "Acidi." "Hydrochloricum" is an adjective (termination -us, -a, -um) agreeing with "Acidum" in gender, number and case. There may even, as in English, be a second adjective in the title, for example "Acidum Hydrochloricum Dilutum," or "Acidum Aceticum Glaciale" ("Glacialis" is an adjective of the third declension nom. -is, -e, gen. -is.) . . . . . The names of salts may again be looked upon as the Latin form of their English names, though not their official English names, for example "Potassium Bromide," "Bromide of Potassium," becomes "Potassii Bromidum." Bromidum the name of the acid constituent is a noun of the second declension as is Potassium but the latter appears in the partitive genitive in the official name. . . . All salts whose names end in -"ide" have names in Latin ending -idum. The names of salts ending in "-ate" have Latin names in a "-as" (gen. -atis) and are masculine nouns of the third declension, example "Sodium Citrate" is "Sodii Citras." Those salt names ending in "-ite" have Latin names masculine and of the third declension in "-is" (gen. -itis), example "Sodium Sulphite, Sodii Sulphis."

The names of alkaloids become in Latin feminine substantives of the first declension with a termination "-ina" (gen. -inæ), example Strychnina. Those of glucosides, bitters and neutral principles are neuter substantitives of the second declension with a termination "-inum" (gen. -ini), example Aloin, Aloinum.

The names of parts of plants may be looked upon also as direct translations, example "Belladonna Leaves, Leaves of Belladonna, Belladonnæ Folia," Folia being a neuter noun in the plural (nom. sing. folium gen. folii pl. folia, gen. foliorum). "Belladonnæ" is the genitive of the feminine noun of the first declension, "Belladonna."

The names of preparations are again similarly formed 'Tincture of Opium, Tinctura Opii"; Tinctura" is a feminine noun of the first declension "Opii" the genitive of the neuter noun of the second declension "Opium."

Some of the cases where the student may find it difficult to understand the Pharmacopœial nomenclature are noted in this paragraph:—Liquor

Ammoniæ, Solution of Ammonia (the hydrate) and hence Spiritus Ammoniæ Aromaticus, Aromatic Spirit of Ammonia, but Liquor Ammonii Acetatis, Solution of the Acetate of Ammonium ($NH_3$); Vinum Antimoniale, Antimonial (adj.) Wine; Liquor Arsenicalis, Arsenical (adj.) Solution; but Liquor Arsenici Hydrochloricus, Hydrochloric (adj) solution of Arsenicum (an old word for Acidum Arseniosum) and Liquor Arsenii Hydrargyri Iodide, solution of the Iodide of Arsenium (the metal) and of Mercury.

The formation of the genitive and plural should as a rule give no trouble but the following nouns have somewhat irregular genitives. Adeps, Adipis, Mel, Mellis; Fel, Fellis; Mas, Maris; Rhizoma, Theobroma, Physostigma, enema, gargarisima, make the genitive in -atis; Aloe, Aloes; Cantharis, Cantharidis; Rhœas, Rhœados; Colocynthis, Colocynthidis; Flos, Floris; Digitialis, Hydrastis, Sinapis do not change in the genitive; Jaborandi, Kino, Catechu, Buchu, Kousso, Peru, Tolu, and most names ending in "l" are indeclinable. Spiritus, Fructus, Cornus, haustus are nouns of the fourth declension with genitives in -ūs.

The gender of Latin substantives may usually be judged by their termination, substantitives in -us and -or being usually masculine (exceptions names of plants in -us, Prunus Virginiana), those in -a are feminine; those in -um and -on and indeclinable nouns neuter.

VOCABULARY of words commonly occuring in the INSCRIPTION. The parts of speech are indicated by the usual abbreviations, as is the gender of the nouns, the case governed by prepositions; the genitive, singular (or plural in the case of plural nouns) and the plural will also be given for substantives and the terminations of the nominative for adjectives, also the accepted abbreviations.

| LATIN. | ABBREVIATION | ENGLISH. |
| --- | --- | --- |
| Acetum—s. neut.-i. -a | | vineger. |
| Acidum—s. neut. -i. -a | acid. or ac. | acid. |
| Ad—prep. acc | | to. |
| Ana—indecl | aa | of each. |
| Aromaticus—adj. -us. -a. -um | aromat | aromatic. |
| Balsamum—s. neut. -i. -a | balsam | balsam. |
| Centimetrum (Cubicum) -i. -a | c. (c.) | centimetre (cubic) |
| Compositus—adj. -us. -a. -um | comp | compound. |
| Congius—s. masc. -i. -i | C | a gallon. |
| Cum—prep. abl | c or cum | with. |
| cum semissee | ss | and a half. |
| Collyrium—s. neut. -i. -a | collyr. | an eye-lotion. |
| Destillatus—adj. -us. -a. -um | dest | distilled. |
| Dilutis—adj. -us. -a. um | dil | dilute |
| Dimidus—adj.-us. -a. -um used | | |
| thus "unciam dimidiam" | ss | a half ounce. |
| Drachma—s. fem.-æ.-æ | dr. or ℨ | a drachm. |
| Extractum—s. neut. -i. -a | extr | an extract |

| Latin | Abbr. | Meaning |
|---|---|---|
| Flavus—adj. -us. a. -um | | yellow. |
| Flexile—adj. -e. -is | | flexible. |
| Fluidus—adjs.-us. -a. -um | fld | (fluid) liquid. |
| Flos—s. masc.-floris.-flores | | a flower. |
| Folium—s. neut. -i. -a. (gen. pl. -orum) | fol | a leaf. |
| Gramma—s. -atis, -ata | g. or gm | a gramme. |
| Granum—s. neut. -i. -a | gr | a grain. |
| Gummi—s. neut. indecl | | a gum. |
| Libra—s. fem. -æ. æ | ℔ | a pound. |
| Lignum—s. neut. -i. -a | | wood. |
| Liquidus—adj. -us. -a. -um | liquid. | liquid. |
| Liquor—s. masc. -oris. -ores | liq | a fluid, a solution. |
| Minimum—s. neut. -i. -a | min. or m | a minim. |
| Mollis—adj. -is. -e | | soft. |
| Mel—s. Mellis pl. Mellita (Mella) | | a honey. |
| Mucilago—s. fem. inis. -ines | mucilag | a mucilage. |
| Niger—adj. nigra.-um | | black. |
| Nux—s. fem. -nucis. -nuces | | a nut. |
| Octarius—s. masc. -i. -i | O | a pint. |
| Oleum—s. neut. -i. -a | ol | an oil. |
| Preparatus—adj. -us. -a. -um | | prepared. |
| Quantum—sufficiat (satis) | q.s. | as much as may be required |
| Radix—s. fem. -icis.-ices | | a root. |
| Recipe—v | ℞ | take. |
| Rectificatus—adj. -us.- a. .um | rect | rectified. |
| Resina—s. fem. -æ. -æ | | a resin. |
| Scrupulus—s. masc. -i. -i | scr. ℈ | a scruple. |
| Semen—s. neut. -inis. -ina (gen. pl. inum) | | a seed. |
| Semis | ss | a half. |
| Spiritus—s. masc. -us. -us | spt | a spirit. |
| Succus—s. masc. -i. -i | | a juice |
| Syrupus—s. masc. -i. -i | syr | a syrup. |
| Tinctura—s. fem. -æ. -æ | tinct | a tincture. |
| Uncia—s. fem. -æ. -æ | oz. or ℥ | an ounce. |
| Viridis—adj. -is. -e | | green. |

Words and phrases commonly occuring in the subscription:—

| Latin | Abbr. | Meaning |
|---|---|---|
| Capsula—s. fem. æ | caps | a capsule. |
| Cataplasma—s. fem. -atis, pl. -ata | | a poultice. |
| Ceratum—s. neut. -i,-a | | a wax ointment, a cerate. |
| Charta—s. fem. -æ. -æ | chart | a paper, (a powder). |
| Collutorium—s. neut. -i. -a | | an eye-wash. |
| Confectio—s. fem. -onis, ones | | a confection. |
| Coque | | boil. |
| Cujus | | of which. |

Detur: . . . . . . . . . . . . . . . . . . . . .det. . . . . . . . . . .let be given.
       detur tales doses. . . . . . . . . . . . . . . . . . . . .let such doses be given.
Divide. . . . . . . . . . . . . . . . . . . . . . . . . . . . . . . . .divide.
       divide in partes æquales. . . . . . . . . . . . . . .divide into equal parts.
       divide in pulveres viginti. . . . . . . . . . . . . . .divide into 20 powders.
Dosis—s. fem. -is. -es. . . . . . . . . . . . . . . . . . .a dose.
Emplastrum—s. neut. -i, -a. . . .empl. . . . . . . . .a plaster.
Emulsio—s. fem.- onis. -ones. . . . . . . . . . . . . .an emulsion.
Enema—s. neut. -atis, -ata. . . . . . . . . . . . . . . .an enema or clyster..
Fiat. . . . . . . . . . . . . . . . . . . . . . . .ft. . . . . . . . . . . .let be made.
       fiat mistura. . . . . . . . . . . . . . .ft. mist. . . . . . .let a mixture be made.
Fiant pulveres viginti. . . . . . . . . . . . . . . . . . . . .let 20 powders be made.
Haustus—s. masc. -us. -us. . . . . . . . . . . . . . . . .a draught. (The entire quantity prescribed to be taken at one dose.)
Lege artis. . . . . . . . . . . . . . . . . . . .l.g. . . . . . . . . . . .according to rule.
Linimentum—s. neut. -i, -a. . . . . . . . . . . . . . . .a liniment.
Lotio—s. fem. onis, -ones. . . . . .lot. . . . . . . . . .a lotion (skin wash).
Massa—s. fem. -æ, -æ . . . . . . . . .mass. . . . . . . .a mass.
       fiat massa et divide in pilulas triginti. . . . . .Make a mass and divide it into 20 pills.
Misce. . . . . . . . . . . . . . . . . . . . .M. . . . . . . . . . . .mix.
Mitte. . . . . . . . . . . . . . . . . . . . . . . . . . . . . . . . . .send.
       mitte doses tales 48. . . . . . . . . . . . . . . . . . . .send 48 doses.
Ne repetatur. . . . . . . . . . . . . . . . n. rep. . . . . . . . .do not repeat.
Numerus—s. masc. -i. -i. . . . . . . . . . . . . . . . . . .a number.
       fiant pilulæ in numero viginti. . . . . . . . . . . .to make 20 pills.
Pilula—s. fem. -æ. -æ . . . , . . . . .pil. . . . . . . . . . .a pill.
       ft. pilula et mitte tales 20. . . . . . . . . . . . . . . .make a pill and send 20 of them.
Pulvis s. fem. -eris. -eres. . . . . . .pulv. . . . . . . . . .a powder.
       ft. pulvis et divide in partes æquales decem. Make a powder and divide it into 10 equal parts.
Signa. . . . . . . . . . . . . . . . . . . . .sig. or "s". . . . . .label.
Signa nomine proprio. . . . . . . . . .sig. n. p. . . . . . .label with its common name.

Words and phrases occuring frequently in the signature.

Ad libitum. . . . . . . . . . . . . . . . . . .ad lib. . . . . . . . . .as much as may be desired.
Alternis diebus. . . . . . . . . . . . . . . .alt. dieb. . . . . . .every other day.
Alternis horis, alterna hora. . . . .alt hor. . . . . . . .every other hour.
Bini—adj. -æ. -a. . . . . . . . . . . . . . . . . . . . . . . . . .two at a time..
Bis in dies. . . . . . . . . . . . . . . . . . . . . . . . . . . . . . .twice daily.
Capiat. . . . . . . . . . . . . . . . . . . . . . . . . . . . . . . . . .take (or let him take).

| Latin | Abbrev. | English |
|---|---|---|
| Cibus. s masc. -i. -i | | food. |
|     post vel ante cibos | p. vel. a.c. | after or before meals. |
|     post cibum vespere | | after the evening neal. |
| Cochleare— s. neut. -is, -ia, | coch | a spoonful. |
| Cochleare Parvum (vel infantis) | | a teaspoonful. |
| Cochleare medium (vel mcdicum) | | a dessertspoonful. |
| Cochleare amplum (vel magnum) | | a tablespoonful. |
| Cyanthus—s. masc. -i. -i | | wine-glass. |
| Diebus tertiis vel quartis | | every third or fourth day. |
| Ex aqua | ex. aq. | in water. |
| Febri durante | | during the fever. |
| Hora—s. fem. -æ. -æ | h. or hor | an hour. |
|     hora decubitus | hor. decub | } at bedtime, on retiring. |
|     hora somni | h.s. | |
|     hora secunda | | at two o'clock. |
|     quaque secunda hora | | } every other hour. |
|     secundis horis | | |
| Indies | | daily. |
| Mane | | in the morning. |
|     primo mane | | on rising. |
|     mane sequente | | the following morning. |
|     mane nocteque | | night and morning. |
| Omni hora | omn. hor | every hour. |
| Omni quadrante hora | | every quarter hour. |
| Per—prep. acc | | by. |
|     per os | | by the mouth. |
| Prandium—s. neut. -i, a- | | dinner. |
| Pro re nata | p.r.n. | as required. |
| Post—prep. acc | | after. |
| Quaque quarta hora | q. q. h | every fourth hour. |
| Quaque sexta hora | | Every sixth hour |
| Quotidie | | daily. |
| Sumat | | let him take. |
| Sumendus—gerundive of sumo | | take. |
| Ter in die | t.i.d | three times a day. |
| Vespere | | in the evening. |

NUMERALS.

1. unus-a-um.
2. duo-æ-o.
3. tres, tria.
4. quattuor.
5. quinque
6. sex
7. septem
8. octo
9. novem
10. decem
11. undecim
12. duodecim.
13. tredecim
14. quattuordecim.
15. quindecim.
16. sedecim
17. septemdecim
18. duodeviginti
19. undeviginti
20. viginti
21. unus et viginti or viginti unus.
24. viginti quattuor.

30. triginta
40. quadraginta.
50. quinquagint
60. sexaginta.
70. septuaginta.
80. octoginta.
90. nonaginta.
100. centum-i.
200. ducenti-æ-a.
1000. mille

ORDINALS.
1st. primus.
2nd. secundus.
3rd. tertius.
4th. quartus.
5th. quintus.
6th. sextus.
10th. decimus.
12th. duodecimus.

ADVERBS.
1. semel (once)
2. bis (twice)
3. ter (thrice).
4. quater.

## CHAPTER IX.

## NOTES ON MAGISTRAL PHARMACY.

To Magistral or Extemporaneous Pharmacy belongs the compounding and dispensing of drugs. Its successful performance naturally has to be preceded by a knowledge of their physical and chemical characters. Dexterity in the art can only be secured by long practice, something for which the medical student has no opportunity. The dispenser stands between the prescriber and the patient and only a very intimate acquaintance with the characters and doses of medicines will enable him to successfully perform his duty to each. The physician who dispenses his own remedies assumes a double liability in that he becomes sponsor for the proper selection of the remedy as well as its preparation so that the patient may take the prescribed quantity without danger to himself.

Carelessness in weighing and measuring medicines should not be tolerated on any account. Guessing at the weight or volume of ingredients is criminal.

**On the Dispensing of Mixtures.** In this procedure as in every other the dispenser must take the greatest care that all the apparatus used must be scrupulously clean. The bottle selected must be of such capacity as to be filled by the ingredients. Sick persons are often so full of fear and doubt that the slightest unusual feature,-insufficient bulk, change in colour, flavour or clearness of a mixture, at once awakens suspicion.

If the prescription presents no incompatibles and the dispenser knows that it will be and remain clear when filled, it may be dispensed directly into the bottle. Or if it is intended that a precipitate is to be produced, the same procedure may usually be followed. It is a good rule to introduce all fluids through a funnel as this serves to keep the neck and sides of the bottle clean and dry. Solutions of common salts and any other solution that is not quite as clear as it ought to be should always be passed through a filter; a little absorbent cotton in the neck of the funnel often serves very well. The fluids of least bulk should as a rule be first dispensed unless there is some special reason for deviating from this, for example very volatile fluids should be dispensed last.

Separate the soluble solids and dissolve them in a portion of the menstruum by trituration in a mortar before placing them in the bottle. Never permit solids that are completely soluble in the vehicle to leave the dispensary undissolved. The method of adding soluble salts directly to the mixture while it may save some time is not to be commended because of the frequency with which their solutions contain foreign matter which requires filtering out. Insoluble dry drugs if prescribed should be reduced to a fine powder, mixed with some of the menstruum and added to the rest. In many instances it is well to suspend insoluble drugs by the addition of gum, mucilage or a viscid fluid such as syrup or glycerin. If the vehicle be water or an aqueous fluid and there are oils, balsams or oleo-resins ordered these should be emulsified before being added to the bottle. The remainder of the menstruum is now added and the bottle corked.

The dispensing of fluid medicines necessitates a more complete acquaintance with the subject of incompatibliity than is the case with any of the other forms of extemporaneous prescriptions.

All solids should be weighed and all fluids measured.

Fluids must be poured from the back of the dispensing bottles so as to save the labels.

Powders must be taken from their bottles with a long spatula.

The physician dispensing his own prescriptions may facilitate his office work by keeping many of the frequently used drugs prepared in concentrated solution. These are made by dissolving a known weight of the drug in a sufficient quantity of the solvent to make a definite volume of the final solution. For instance if eight drachms of Bromide of Potassium are dissolved in the quantity of water required to make a solution measuring four fluid ounces, each four fluid drachms of the latter will then contain one drachm by weight of the Bromide. These are called dispensing solutions and are quite different to the percentage solutions of the chemical laboratory and to those of the Pharmacopœia. Such salts as the Bromides, Iron Tartarate, Magnesium Sulphate, Potassium Iodide, Chlorate and Nitrate, may readily be kept in this form; Bicarbonates in solution are liable to change as do many of the organic preparations such as Chloral. Dispensing solutions should always be kept in the dark. Though the result is hardly so good as when the Pharmacopœial method is used, flavoured syrups may often be prepared by adding to a Simple Syrup a liquor of the flavour required. In many other cases concentrated liquors or extracts may be obtained from the pharmaceutical houses which may be diluted so as to approximate the official preparations. In some cases it is of advantage to keep weighed out and wrapped as powders, the insoluble drugs in powder form needed for such a mixture as Mistura Cretæ. Infusions especially of digitalis should always be freshly made.

**Percentage Solutions.** The disadvantages of the Imperial System are most clearly seen when dispensing this class of solutions. Absolute exactitude cannot be attained. For dilute solutions (1-3%) of highly active drugs, e.g. Strychnine, Atropine, the following method is to be recommended:—110 minims of water weigh 100 grains; therefore if a 2% solution be required 2 grains are dissolved in 110 minims. The bulk of the resulting solution is slightly greater than 110 min. and its weight exceeds

100 grains. but for all practical purposes 1 min. will contain 2/100 grain. If one ounce were required, the simplest procedure is to make 4½ x 110 or 495 min. and throw away 15. When a more concentrated solution e.g. an ounce of a 10% solution of a salt, is prescribed a somewhat more complex procedure may best be followed, as it is impossible to estimate what the specific gravity of the resulting fluid will be. The following procedure is often pursued:—There are 437 grains to the fluid ounce, 10% of this is 43. If 43 gr. salt were dissolved in 394 min. the bulk would not be 1 fl. oz., but it 50 gr salt were dissolved in 450 of water the bulk would obviously exceed a fluid ounce and in consequence such a solution might be made and the excess thrown away. Very frequently and with sufficient accuracy for many physician's purposes, percentage solutions are not made as above by weight but by weighing the solids and measuring the liquids, one minim being then considered as weighing one grain; one ounce, one fluid ounce.

**Emulsions.** These are mixtures of resinous or oily substances with water. They consist of minute particles of the active substance surrounded with, kept apart, and in suspension by means of mucilage made from one of the gums. Acacia or Tragacanth are commonly selected in the dispensary. Perfect natural emulsions are to be seen in milk and in the yolk of egg.

Of the resinous drugs Asafetida, Myrrh, Copaiba, Extract of Male Fern, the Tinctures of Cannabis Indica, Tolu, the Compound Tinctures of Guaiacum, and Benzoin, frequently require treatment, as do Cod-Liver and Castor Oils, Turpentine and Camphor. In the case of the gum-resins such as Asafetida which contains a good deal of gummy matter it is not necessary to add extraneous gum to obtain an emulsion, that which is part of the drug being sufficient on trituration with water.

Emulsions are prepared with the aid of a mortar and flat pestle. A thick mucilage is first made and with constant stirring a portion of the drug is added in small quantities until the emulsion is obtained, when the balance is added alternately with the remaining water in successive portions until the whole is emulsified.

With oils a second method may be adopted, called the English Method. Two or three parts by weight of Gum Acacia are triturated in a mortar with eight parts of Oil until the gum is completely suspended. Then one and a half parts of Water are added at once when a few revolutions of the pestle will secure an emulsion. The balance of the Water is now to be added in successive quantities until the whole is used. If the emulsion is not completed in the first stage of the process or the water is added too freely the oil separates and the emulsion is said to "crack" and cannot be restored.

**On the Dispensing of Pills.** The prerequisite of a properly made pill is a proper pill mass. This should possess the following characters, consistence, cohesiveness, and plasticity. Proper consistence is essential for if too hard the mass may not be divided into pills while if too soft the pills made will not retain their shape and will tend to run together.

In dispensing a pill, the ingredient present in smallest amount and especially if it is very potent, is first tritrated in the mortar with a gradually increasing quantity of one of the other ingredients. All the other ingredients are ground together to a thoroughly smooth impalpable powder before the excipient is added. After the excipient has been thoroughly ground up and mixed with other ingredients (the mortar and pestle should be scraped down several times with a stiff-bladed metal spatula during the process) the mass is scraped together and transferred to the pill machine or is rolled in the hands into a smooth ball, then on the pill tile into a pipe, which must be kept of the same bore throughout its length. By constant rolling the pipe (after its ends have been squared by pressure) is made to reach the length indicated on the tile for the required number of pills. It is then cut into pill lengths as shown on the tile and each length carefully rounded on the hand or tile with the finger tips. When all the pills are rounded they may be finished by placing them under the pill finisher with a little dusting powder. The finisher is made to describe a figure-of-eight movement until the pills are round.

Pills must be round, not cracked, and not sticky. If stickiness develops during rolling, a dusting powder, starch, talc, or powdered liquorice, may be used on hands and tile.

Excipients must be added slowly and only as much as is needed. Fluid excipients should be dropped first on the spatula and added only one drop at a time.

The following are some of the most useful excipients:—

WATER.—Of use where there are considerable proportions of aqueous extracts as those of Aloes, or Cascara; where there is a gummy substance as Asafetida, or with those holding Hard Soap.

GLYCERINE OF TRAGACANTH.—One of the best for general use, being powerfully adhesive, at the same time preserving the consistence of the pill and promoting its solution. Tragacanth has large powers for absorbing water.

SYRUP OF GLUCOSE.—Much used in the official pills and particularly where it is not necessary to confer much adhesiveness to the mass.

EXTRACT OF MALT.—Makes a good excipient for general use, not being eligible of course in those pills where vegetable substances are to be avoided as with pills of Silver Nitrate.

POWDERED LICORICE ROOT AND POWDERED EXTRACT OF LICORICE.—These possess mild adhesiveness. The former because of its absorbent power is useful with very soft masses, it also makes an excellent dusting powder for the finished pills.

POWDERED HARD AND CURD SOAP.—The former is of use in making those pills containing vegetable substances as powdered crude drugs, the extracts, and the gum resins such as Myrrh or Asafetida. The latter is especially helpful with pills of Creosote or the Essential Oils. Avoid using soap for massing metallic salts, acids, or compounds of Tannin.

KOALIN.—Of use in massing easily combustible substances such as Permanganate of Potassium. Nitrate of Silver and Phosphorus. Cohesion is secured by the addition of a fatty substance such as Resin Ointment.

POWDERED ACACIA.—Is mentioned only that it may be avoided unless combined with some fibrous powder as Licorice or Althaea Powder. Pills made with Acacia are apt to become extremely hard and have been known to pass through the bowel undissolved.

COATING OF PILLS.—For the physician to attempt anything more than a simple dusting of new made pills with some inert dry powder such as Licorice would be to tempt disaster as the process of coating with sugar, silver, or gelatin, other than using a gelatin capsule, belongs to the expert dispenser. Pills intended for solution in the bowel may be coated with a preparation of Keratin in which case they must be made with a fatty excipient, and are difficult to make. They also may be coated with melted Salol which is placed in a shallow container and the pills rotated in it until covered. If Salol is used the excipient must not be made of fat.

**On the Dispensing of Capsules.**—For this purpose the drugs used are powdered finely and placed in the capsule by the aid of a spatula or of a patent capsule filler after being accurately subdivided. The patent filler consists of a stand which supports the capsules in an upright position and a sliding funnel, riding over the base, through which the powder is poured into each capsule. Capsules are made of several sizes, holding from one to ten grains of powdered quinine and more of the denser drugs. All drugs should be reduced to powder before being dispensed in this way.

A second method is to proceed in the same manner as in the making of pills up to the point of the division of the mass when the sections instead of being moulded into pills are inserted into the capsule after being rolled to the proper diameter.

Oils, Balsams, and Alcoholic Solutions may be dispensed in this way but care must be taken to seal the cover on by moistening the base of the capsule with a brush dipped in water at the part which is covered by the lid before this is placed in position. This effectually prevents the contents finding their way out and air from entering. Aqueous fluids may not be given in this manner unless administered at once. Soft capsules are made and filled by the large manufacturers and are not readily dispensed by hand unless special apparatus is available.

In dispensing capsules by hand the skin must be perfectly dry otherwise the fingers will soften the outside of the capsule to which any powders will adhere, making an unsightly product and giving their unpleasant taste to the gelatin. The filled capsule should still possess its lustre and be quite free of the taste of the enclosed medicine.

**On the Dispensing of Cachets.**—This is perhaps the most elegants way of administering powders of moderate bulk, it being possible to enclose about double as much as by capsule. As in the dispensing of dry powders by capsule it is first necessary to convert everything to powder form. To turn out cachets properly requires the use of a cachet machine though a serviceable substitute may be made by using two bottles having wide mouths of sufficient inside diameter to hold a half-cachet. The powder is placed in one half being careful not to allow any to fall upon the projecting edge. The edge of the other half is now moistened with a brush dipped in water, and a very little having been applied the empty half is inverted over the other and with the application of slight pressure becomes adherent. The use of the machine permits the same procedure to be accomplished much more rapidly. Fluids and deliquescent drugs may not be dispensed in this manner. As there are several sizes of cachets available that best suited to the bulk of the medicine should be selected.

**On the Dispensing of Powders.**—Drugs selected for dispensing in powders are commonly those with little unpleasant taste. As we have seen nauseous powders are best given in capsules or cachets. Deliquescent drugs or those affected deleteriously by the atmosphere should not be dispensed unless wrapped in oiled paper.

Every remedy should be reduced to fine powder, and if several are to be mixed this is to be done in the usual order, beginning with those of smallest bulk and gradually adding those which are larger. Powders may be triturated in a mortar with the pestle if light trituration is used:— hard pressure is apt to cause caking making the resulting powder difficult to swallow.

A very useful way to obtain the thorough admixture of powders is to pass them repeatedly through a fine sieve. If the total quantity is small, powders may be readily and well mixed by triturating them together upon a piece of paper with a spatula and then passing them once through a sieve. The division of powders may be done with the spatula, equality in size being determined by the aid of the eye, or more exactly each powder may be weighed.

Powder papers should be of equal size and when folded of the same width and length, this being determined by the size of the box in which they are to be placed. The folding over of the ends should be the same in each so as to secure absolute uniformity.

**On Suppositories, Bougies, Pessaries.**—The active agent is reduced to a powder or to a paste and incorporated with the Cocoa Butter which has been melted at a low temperature (preferably on a water bath). When at the point of congealing and while still possible to pour the mixture it is run into metal moulds which have been previously cooled on ice and moistened with Soap Liniment, or a fixed oil such as Almond or Olive Oil. The mould is again placed on the ice until the product has become solid when the suppositories are removed and may then be placed in impervious boxes or those lined with either tin foil or paraffined paper.

Suppositories may likewise be made by hand, by allowing the mixture to become cooled to that point where it is plastic but not hard, when the mass is rapidly moulded on a pill tile into conical shapes of definite weight. A third method is to make paper cones which having been oiled are placed, open end up, in sand or linseed meal. The melted mass is now poured into the cones and the vessel containing them is set aside in a cool place. When solid the suppositories are removed from the paper holders and boxed.

In a large way they are manufactured by a special machine which by pressure forces the mixed ingredients, prepared in the cold, into moulds of such shape as may suit the need of the prescriber. This is called the "Cold Method."

Pessaries, Bougies and some Rectal Suppositories are best made with a gelatin base, from a mixture of gelatin and glycerin. This is not useful in the case of Tannic and Carbolic Acids nor with Ichthyol.

Essential Oils, as Oil of Eucalyptus, are best made up with the addition of a small quantity of white wax to the Cacao Butter. About the same weight of wax as of the oil is necessry to make a firm suppository. Wax may be added also in very warm weather.

Heavy Salts such as Acetate of Lead tend when the suppository is made by heat to gravitate during the cooling to the apex where it forms a hard brittle mass. For these the method of making by hand is perhaps the most useful.

**On the Dispensing of Ointments.**—With no other group of preparations is it so essential to have all solids reduced to a fine powder as with ointments, unless perchance they are readily soluble in fats. The subdivision should be so fine that when incorporated with the base no grittiness is evident to either the eye or the finger.

Ointments may be made upon an ointment slab, (the reverse of a pill tile may be used or a square of ground glass), or with mortar and pestle. The active drug is first made into a paste with a few drops of water, spirit or glycerin. It is then triturated with a small quantity of the base until thoroughly mixed. The rest of the base is then added and the trituration continued until the whole is incorporated. When completed the ointment is to be dispensed in a box or jar. This should be done cautiously so as not to smear the outside of the container and so as to leave a smooth finished surface to the ointment itself. The spatula aided by the flame of a gas or alcohol lamp over which the inverted jar is held for a moment will suffice for this.

The base selected for any ointment should be such as will fulfil the purpose of the prescription, some fats being absorbed by the skin, others not. It should be chosen with a view to avoiding chemical reaction between it and the active constituent. As already stated the chief bases are Wool Fat, Lard, and Paraffin. Their absorption by the skin and their power of absorbing liquids is in the order of mention. For extemporaneous prescriptions Wool-Fat is much more used than in the making of the official ointments.

For impressing the general system then the base should be Wool-Fat or Lard, preferably the former. This used alone makes a rather stiff ointment which is difficult to prepare and to apply. This may be avoided by the addition of a small proportion of Lard or Olive Oil.

For those to be used purely for their local effect, Soft Paraffin or a mixture of Hard and Soft, depending upon the climate, makes the ideal preparation; in cold weather less, in warm more of the Hard Paraffin is used.

The Pharmacopœia directs the dispenser to use Yellow Paraffin if colored drugs are to be dispensed and the White for those that are colourless. This is a good rule for all but those ointments to be applied to the eye. White Paraffin is made by bleaching the Yellow with the aid of the mineral acids and there is likely to be a trace of acid present which makes it unsuited for application to the conjunctiva. For these unguents use the Yellow Paraffins. The greatest precaution to obtain ointments absolutely free of grit should be taken when for use in the eye.

**On the Dispensing of Plasters.**—The making of plasters has been so completely passed over to the manufacturing pharmacist that it seems needless to discuss the subject. Almost any formula can be had already spread by machinery with such art that the unskilled hand may not hope to obtain such perfect results from a pharmaceutical, let alone from the therapeutical, standpoint.

BOTTLES.—Those used for dispensing may be had in plain or variously coloured glass and of oval, round, or square shape. Medicines for internal use are commonly dispensed in flint or colourless bottles which are somewhat more expensive than bottles made of green glass but the better appearance makes ample return for the additional cost. Amber and blue glass bottles are in frequent use for sending out poisons and also for storing solutions which may be affected deleteriously by actinic light. Vials, for poisons, of unusual shape and studded with raised points of glass so that they may be instantly recognized. even in the dark, are advised. Prescription bottles vary in size from those containing a drachm to those holding as much as a pint or more. Cylindrical flint glass bottles holding one, two and four fluid drachms are known as homeopathic vials, and are of use in the dispensing of small quantities of eye lotions and other remedies to be administered in minute doses. For ordinary prescriptions bottles are made to contain ½, 1, 2, 3, 4, 6, 8, 10, 12 and 16 ounces. After the three ounce there are no odd sizes made for dispensing so that prescriptions calling for more than that quantity of fluid should be written for even numbers of ounces.

LABELS.—These should be of a style and shape to suit the special package to which they are attached. It is well to have two sizes for bottles. They should bear the physician's name and address, and if desired his office hours and telephone number. These items should be printed or lithographed plainly but unobtrusively, so as to leave ample space for directions to the patient. They may be already gummed if that is wished, though labels so prepared are apt to adhere in warm weather and thus become spoiled.

Labels ought to be attached so as to make the most symmetrical parcel possible, neither close to the top nor to the bottom of a bottle, but rather over the middle third of its face.

CORKS for dispensing bottles should be of the longer cuts, should be kept in a moist atmosphere to prevent their becoming friable and the one used should be of such size as not to require insertion for more than half its length.

BOXES FOR POWDERS.—These are made of paper and are oblong, or square in shape. They may be of the well-known telescope design or have the lid lift from the base, these being the more costly. The upper surface of the cover is reserved for the label.

BOXES FOR PILLS.—Made of paper and ordinarily flat and circular in shape.

BOXES AND JARS FOR OINTMENTS.—These may be of wood, paper, tin or glass. The two former kinds are made impervious by preparatory treatment with a solution of silica. Glass jars may have covers of the same material, or of metal which ought to be non-corrosive. These containers are spoken of as being of ½, 1, 2, 3, 4, 6, and 8 ounce in size, as determined by the capacity of each. Those made of glass are preferable but are the most expensive. Labels are commonly applied to the upper surface of the lids but in the case of those having metal covers it may be found difficult without a special mucilage to keep them adherent. With the glass jars having metal covers this may be obviated by placing the label either upon the side or the bottom.

## ADDENDUM.

Extract from the Report of the Council of the British Medical Association on the Adoption of the Metric System of Weights and Measures by Medical Practitioners in Dispensing and Prescribing.

British Medical Journal Suppl.. 1911, 1 p. 205

### A.—Transitional Procedure Suggested for Adoption by Medical Practitioners.

(5) To practitioners who have been trained according to the present system, the Council recommends the adoption of transitional procedure, which would enable them at once to adapt their prescriptions to the measures of the Metric System, and so avoid the drawbacks that would arise from a divergence in practice between junior and senior practitioners, and would also at once secure for senior practitioners the advantages which make the general adoption of the Metric System desirable.

(6) The difficulty before the practitioner who has been trained to think in terms of grains and minims is to translate his quantities readily into grams and cubic centimetres, and if absolute exactitude were necessary he would require the constant use of tables of equivalents. In practice, however, the most common mode of administering medicines is by spoonfulls, and even when these are poured carefully into a medicine glass the range of variation is relatively wide and the dosage must be such as to make this variation entirely safe. The Council, therefore, feels justified in recommending to the profession, as a transitional measure the following methods which are based on the actual conditions of British practice, and for the suggestion of which the Council is indebted to Dr. R. C. Buist. These will be found to give automatically the conversion of a dosage in grains and minims into a prescription which the dispenser can measure in grams and cubic centimetres with an approximate exactitude well within the range of variation of spoon measures.

*Mixtures.*

(7) In the prescription of an 8 oz. mixture, of which each tablespoonful is to contain
    (a) Tr. Belladonnæ, m V.
       Spt. Ætheris, m X.
       Vin. Ipecac., m XV.
       Syr. Scillæ, m XX.
       Inf. Senegæ ad ½ oz. (*i.e.*, m CCXL).
The Metric prescription for the mixture would be
    (b) Tr. Belladonnæ, 5.
       Spt. Ætheris, 10
       Vin. Ipecac., 15.
       Syr. Scillæ, 20.
       Inf. Senegæ ad 240.

On comparing (*a*) and (*b*) it is evident that the numbers are the same in both.

(8) The prescriber intends a mixture to contain certain substances in fixed portions, which will be the same in the single dose and in the bulk, and will not be affected, whether the measures be stated in minims or in cubic centimetres; the numbers of minims will be larger, but the proportions will be the same. The exact factors for the conversion of grams into grains and of cubic centimetres into minims are 15,4324 and 16,906, respectively. The procedure used in the above example is to take 16 as a near approximation to each of these numbers. (The extent to which this is inexact may be stated as 4 drops in a teaspoonful.) Now, in ordinary prescribing, 16 doses is the most common of all orders, as represented by tablespoonful doses of an 8-oz. mixture. If, therefore, in such a mixture the prescriber orders the numbers of minims of the drugs A, B, C, D, E in each tablespoonful which he would order in a prescription in English measures, but omits the symbols, and if the dispenser measures in each case the same numbers of cubic centimetres into the bottle, the conversion from English into Metric measures will be automatically completed. Thus it is recommended that the practitioner who wishes to write a prescription for Metric measures should simply write without symbols the drugs with the number of grains or minims he intends to give in each spoonful, and that the dispenser be instructed that each prescription where no symbols are written are to be dispensed in Metric measures.

For teaspoonful doses the bulk would be 2 ozs. or 60 c.c. and for dessertspoonful doses, 4 ozs. or 120 c.c.

(9) The following prescriptions are given in illustration:

(*a*) Recipe—
    Tr. Nucis Vom., 5.
    Inf. Quass. conc. ad 60.
    Sig. Teaspoonful in water before each meal.

(*b*) Recipe—
    Tr. Digitalis, 7.5.
    Spt. Ætheris, 10.
    Dec. Scoparii ad 120.
    Sig. Dessertspoonful morn. and night.

(*c*) Recipe—
    Ac. Hydrocyan, dil., 3.
    Liq. Morph. Mur., 10.
    Syr. Tolut., 30.
    Inf. Rosæ Acid. ad 240.
    Sig. Tablespoonful thrice daily.

## Solutions.

(10) In ordering solutions for various purposes the proportions are so evident that no difficulty arises, and the only point to be borne in mind is the total quantity desired. Thus—

(a) Cocain Hydrochlor., 3
Aq. ad 60.
Sig 5 °/₀ Cocain Hydrochlor.

(b)°Argent. Nitrat., 1.
Aq. destil, 50.
Sig. 2% Silver Nitrate.

## Pills and Powders.

(11) The procedure in ordering pills and powders must be somewhat different from that hitherto described. The order for a pill or powder is based on fractions or small multiples of the grain. The prescriber should therefore become familiar with the equivalence 1 grain = 0.06 gram, which is sufficiently exact for practical purposes. To facilitate the work of the dispenser the number of pills or of powders ordered should be a multiple of ten. Thus—

Recipe—
Aloin.
Podophylli Resinæ.
Jalapæ Resinæ.
Ext. Hyoscyami aa 0.015.   M. ft. pil.
M. 10.
Sig. One after each meal.

For his pill mass the dispenser simply shifts the decimal point of the prescription.

## Linear Measures.

(12) The equivalence 1 inch = 2.5 c.m is used in practice.

## SUMMARY.

(13) The procedure here recommended for the use of medical practitioners may thus be summarized:

(a) The prescription is still to be based on the single dose (*b*) In the case of mixtures 16 doses are to be ordered by writing with figures only the number of grains or minims of each ingredient in one spoonful.

(*c*) In the case of pills and powders 10 are to be ordered and the prescription is to give jn figures only the metric equivalent of the grains of each ingredient in the single dose.

(*d*) The dispenser is to be informed that every prescription written without symbols is to be dispensed in Metric measures.

(14) The adoption of the foregoing suggestions would overcome the difficulty of introduction of the new system by a medical practitioner who does his own dispensing, or by one whose dispensing is usually done by the same chemist. For such cases no intervention by the Divisions will be necessary beyond that of bringing this Report under the notice of the local profession. It can be left to each practitioner to take his own course.

# INDEX.

The official Latin names of the drugs of the materia medica are not included in the index as they can readily be found in Chapters V and VI.

| | | | |
|---|---|---|---|
| Acacia | 26 | Bile | 58 |
| Aceta | 11 | Bismuth | 41 |
| Acetic Acids | 26 | Black Draught | 92 |
| Acetanilide | 26 | Black Wash | 66 |
| Acetic Ether | 31 | Blaud's Pills | 59 |
| Acetomorphine Hydrochloride | 102 | Blistering Liquid | 46 |
| Acids | 26 | Blue Pill | 65 |
| Aconite | 29 | Boracic Acid | 42 |
| Adrenine | 102 | Borax | 42 |
| Adrenalin | 102 | Boric Acid | 42 |
| Alcohol | 31 | Boroglycerine | 103 |
| Alkaloids | 10 | Bougies | 14, 111, 131 |
| Almonds | 34 | Brandy | 31 |
| Aloes | 31 | Bromides | 85, 94, 103 |
| Aloin | 32 | Broom | 92 |
| Aluminium | 32 | Buchu | 43 |
| Alum | 32 | Burgundy Pitch | 83 |
| Ammonia | 33 | Butyl-chloral | 43 |
| Ammoniacum | 33 | | |
| Ammonium | 33 | Cacao Butter | 78 |
| Ammonium Bromide | 103 | Cachets | 14, 114, 130 |
| Amyl Nitrite | 35 | Cade, Oil of | 75 |
| Anise | 35 | Caffeine | 43 |
| Antifebrin | 26 | Cajuput | 75 |
| Antimony | 35 | Calabar Bean | 82 |
| Antipyrine | 81 | Calcium | 43 |
| Antitoxines | 106, 107 | Calcium Lactate | 103 |
| Aquæ | 11 | Calomel | 65 |
| Arbutin | 100 | Calumba | 45 |
| Argyrol | 103 | Calx | 44 |
| Arnica | 37 | Camphor | 45 |
| Arsenic | 37 | Canada Balsam | 98 |
| Asafetida | 38 | Canada Turpentine | 98 |
| Aspirin, see Acetyl Salicylicum | 102 | Cantharides | 46 |
| Atropine | 40 | Caoutchouc | 47 |
| | | Capsicum | 47 |
| Balsams | 9 | Capsules | 15, 114, 129 |
| Balsam of Peru | 39 | Caraway | 47 |
| Balsam of Tolu | 39 | Carbon | 47 |
| Bases for Lozenges | 100 | Carbon Bisulphide | 47 |
| Bases for Pills | 128 | Carbolic Acid | 26 |
| Bearberry | 100 | Cardamons | 47 |
| Beeswax | 49 | Caron Oil | 44 |
| Belladonna | 39 | Cascara | 48 |
| Benzoic Acid | 41 | Cascarilla | 48 |
| Benzoin | 41 | Cassia | 48 |
| Benzol | 41 | Castor Oil | 77 |
| Beta-naphthol | 74 | Cataplasmata | 15, 103 |

| | | | | |
|---|---|---|---|---|
| Catechu | | 49 | Dessication | 11 |
| Caustic, Lunar | | 37 | Digitalis | 56 |
| Caustic, Potash | | 85 | Dill | 35 |
| Cereates | 15, | 103 | Dionin, see Ethyl morphine | 104 |
| Cerium | | 49 | Discs | 13 |
| Chalk | | 43 | Distillation | 11 |
| Charcoal | | 47 | Diuretin, see Theobromina | 106 |
| Charta | 11, | 49 | Domestic Measures | 8 |
| Cherry-laurel | | 70 | Dosage for Children and the | |
| Chiretta | | 49 | Aged | 18 |
| Chloral | | 50 | Donovan's Solution | 38 |
| Chloroform | | 50 | Dover's Powders | 79 |
| Chromic Acid | | 27 | | |
| Chromic Anhydride | | 27 | Elaterin | 56 |
| Chrysarobin | | 36 | Elder Flowers | 90 |
| Cimicifuga | | 50 | Elixers | 15, 104 |
| Cinchona | | 51 | Emplastra | 12, 57, 132 |
| Cinchonidine | | 51 | Emulsions | 15, 104, 127 |
| Cinchonine | | 51 | Enemata | 15, 110 |
| Citric Acid | | 27 | Epsom Salts | 73 |
| Citrine Ointment | | 66 | Ergot | 57 |
| Cloves | | 48 | Ergotoxine | 57, 104 |
| Clysters | 15, | 110 | Eserine | 82 |
| Coal Tar | | 83 | Ether | 30 |
| Coca | | 52 | Ethyl Chloride | 104 |
| Cocaine | | 52 | Ethylmorphine Hydrochloride | 104 |
| Cochineal | | 53 | Ethyl Nitrite | 71 |
| Codeine | | 80 | Ethylate of Sodium | 72 |
| Cod-liver Oil | | 77 | Eucalyptus | 57 |
| Colchicum | | 53 | Eucalyptus, Oil of | 76 |
| Colchicine | | 53 | Euonymus | 58 |
| Cold Cream | | 103 | Eucalyptol | 105 |
| Collodia | 11, | 87 | Expression | 10 |
| Collodion | 11, | 87 | Extraction | 10 |
| Collodium | | 87 | Extracts | 12, 58 |
| Collyria | 111, | 15 | Eye-washes | 15, 111 |
| Colocynth | | 54 | | |
| Compound Spirit of Ether | | 30 | Fennel | 61 |
| Confections | 12, | 54 | Figs | 60 |
| Coniine | | 54 | Filtration | 11 |
| Conium | | 54 | Formaldehyde | 105 |
| Copaiba | | 54 | Fowler's Solution | 38 |
| Copper | | 56 | Frankincense | 99 |
| Coriander | | 55 | Friar's Balsam | 41 |
| Corrosive Sublimate | | 65 | Fumigations | 15 |
| Cotton | | 63 | | |
| Cotton, Absorbent | | 103 | Galbanum | 61 |
| Cream of Tartar | | 86 | Galenical Pharmacy | 6 |
| Creosote | | 55 | Gallic Acid | 27 |
| Cresol | | 102 | Galls | 61 |
| Cresylic Acid | | 102 | Gamboge | 45 |
| Croton Oil | | 76 | Gargles | 105, 111 |
| Cubebs | | 55 | Gargarismata | 105 |
| Curd Soap, see Sapo Animalis | | 90 | Gelatine | 61 |
| Cusparia | | 56 | Gelsemium | 61 |
| Cusso | | 56 | Gentian | 61 |
| | | | Ginger | 101 |
| | | | Gloniin | 72 |
| Dandelion Root | | 98 | Glucosides | 10 |
| Decoctions | 12, | 56 | Gluside | 62 |

| | |
|---|---|
| Glycerin | 62 |
| Goa Powder | 36 |
| Goulard's Extract | 84 |
| Goulard's Water | 84 |
| Gregory's Powder | 88 |
| Grey Powder | 65 |
| Griffith's Mixture | 59 |
| Guaiacol | 105 |
| Guaiacol Carbonate | 105 |
| Guaiacum | 63 |
| Gums | 9 |
| Guy's Pill | 91 |
| | |
| Hamamelis | 63 |
| Hemidesmus | 64 |
| Hemp, Indian | 46 |
| Henbane | 67 |
| Heroin, see Aceto morphine | 102 |
| Hexamethylene-tetramine | 105 |
| Hoffman's Anodyne | 30 |
| Honey | 13, 73 |
| Honey; acidulated | 13, 73 |
| Homatropine | 64 |
| Hops | 72 |
| Horse-radish | 37 |
| Hydrobromic Acid | 27 |
| Hydrochloric Acid | 27 |
| Hydrocyanic Acid | 28 |
| Hydrogen Peroxide | 71 |
| Hydrous Wool Fat | 30 |
| Hyoscine | 67 |
| Hyoscyamine | 67 |
| Hyoscyamus | 67 |
| Hypodermic Injections | 12, 67 |
| Humulus | 72 |
| | |
| Idiosyncrasy | 17 |
| Imperial System | 6, 7 |
| Incompatibility | 20 |
| Incompatibility Chemical | 20 |
| Incompatiblilty Pharmaceutical | 22 |
| Incompatibility Pharmacological | 20 |
| Incompatibility Therapeutical | 20 |
| Indian Hemp | 46 |
| India Rubber | 47 |
| Infusion 10; Infusions | 12, 67 |
| Injections, Hypodermic | 12, 67 |
| Inscription | 116 |
| Iodine | 68 |
| Iodoform | 67 |
| Ipecacuanha | 68 |
| Iron | 58 |
| | |
| Jaborandi | 69 |
| Jalap | 69 |
| James' Powder | 35 |
| Juices | 13, 97 |
| Juniper, Oil of | 76 |
| Juniper Tar Oil | 75 |

| | |
|---|---|
| Kaolin | 69 |
| Kino | 69 |
| Konseals | 14, 114, 130 |
| Kousso | 56 |
| Krameria | 70 |
| | |
| Lactic Acid | 28 |
| Lactose | 89 |
| Lamellæ | 13 |
| Lanolin | 30 |
| Lard | 30 |
| Laudanum | 79 |
| Lavender, Oil of | 76 |
| Lead | 83 |
| Leeches | 64 |
| Lemon | 70 |
| Lime | 44 |
| Liniments | 71 |
| Linseed | 71 |
| Liquors | 13, 71 |
| Liquors Concentrated | 13, 71 |
| Liquorice | 62 |
| Lithium | 72 |
| Litharge | 84 |
| Liver of Sulphur | 85 |
| Lobelia | 72 |
| Logwood | 63 |
| Lotions | 13, 72 |
| Lozenges | 14, 100 |
| Lupulin | 72 |
| Lupulus | 72 |
| | |
| Maceration | 10 |
| Magistral Pharmacy | 6 |
| Magnesium | 73 |
| Male Fern | 60 |
| Marc | 10 |
| Materia Medica | 5 |
| Measures, 5; Imperial, 6; Metric, 7; Domestic | 8 |
| Meconic Acid | 78 |
| Mel | 73 |
| Menthol | 73 |
| Mercury | 64 |
| Methylene Blue, see Methylthionina | 105 |
| Metric System | 7, 8 |
| Mezereon | 74 |
| Minderus' Spirit | 34 |
| Milk Sugar | 89 |
| Mitigated Caustic | 37 |
| Mixtures | 13, 74, 112, 125 |
| Morphine | 79 |
| Mucilages | 13, 74 |
| Musk | 74 |
| Mustard | 93 |
| Myrrh | 74 |
| Naphthol | 74 |
| Nitre | 86 |
| Nitrites, see Amyl, Ethyl, Sodium | |

| | |
|---|---|
| Nitrous Ether | 96 |
| Nitroglycerine | 72, 98 |
| Nitric Acid | 28 |
| Nitro-hydrochloric Acid | 28 |
| Nutmeg | 74 |
| | |
| Official | 5 |
| Officinal | 5 |
| Oils, fixed 9; Volatile, 9; Essential, 9; see also Olea. | |
| Ointments | 14, 100, 109, 131 |
| Oleo-resins | 9 |
| Oleic Acid | 28 |
| Olive Oil | 77 |
| Opium | 78 |
| Orange Peel | 38, 39 |
| Orange-flower Water | 39 |
| Otto of Roses | 89 |
| Ox Bile | 58 |
| Oxymel | 73 |
| | |
| Panama Bark | 87 |
| Pancreatic Solution | 72 |
| Papers | 11, 49 |
| Paraffin | 80 |
| Paraldehyde | 81 |
| Paregoric | 79 |
| Pareira | 81 |
| Pelletierine | 63, 106 |
| Pepper | 83 |
| Peppermint, Oil of | 76 |
| Pepsin | 81 |
| Percolation | 10 |
| Peru Balsam | 39 |
| Pessaries | 14, 110 |
| Pharmacology | 6 |
| Pharmacodynamics | 6 |
| Pharmacognosy | 5 |
| Pharmacy | 5 |
| Pharmacopœia | 5 |
| Phenacetin | 81 |
| Phenozone | 81 |
| Phenol | 26 |
| Phenolphthalein | 106 |
| Phosphoric Acid | 28 |
| Phosphorus | 81 |
| Physostigma | 82 |
| Physostigmine | 82 |
| Picrotoxine | 82 |
| Pilocarpine | 69 |
| Pills | 13, 82, 113, 127 |
| Pimento | 82 |
| Plasters | 12; 57, 132 |
| Plummer's Pill | 66 |
| Podophyllin | 84 |
| Pomegranate | 63 |
| Poppy | 80 |
| Posology | 6, 17 |
| Potassium | 85 |
| Poultices | 15, 103 |
| Powders | 13, 86, 114, 130 |
| Prepared Coal Tar | 83 |
| Prescribing | 108 |
| Prescription-writing | 115 |
| Prunes | 86 |
| Pulverisation | 11 |
| Pyrethrum | 87 |
| Pyroxylin | 87 |
| | |
| Quassia | 87 |
| Quillaia | 87 |
| Quinine | 51 |
| | |
| Rectified Spirit | 31 |
| Red Sanders Wood | 86 |
| Red Poppy Petals | 88 |
| Resin | 87 |
| Rhatany Root | 70 |
| Rhubarb | 88 |
| Rochelle Salt | 94 |
| Roses | 88 |
| Rosemary, Oil of | 77 |
| Saccharin | 62 |
| Saffron | 55 |
| Salacetic Acid | 102 |
| Salicin | 89 |
| Salol | 89 |
| Salicylic Acid | 89 |
| Saltpetre | 86 |
| Salvarsan | 104 |
| Sandal Wood, Oil of | 77 |
| Santoin | 90 |
| Sarsaparilla | 90 |
| Sassafras | 91 |
| Scammony | 91 |
| Scotch Paregoric | 79 |
| Scott's Dressing | 65 |
| Seidlitz Powders | 94 |
| Senega | 92 |
| Senna | 92 |
| Sera | 15, 106 |
| Serpentary | 93 |
| Sherry | 31 |
| Signature | 116 |
| Silver | 37 |
| Soap | 90 |
| Sodium | 94 |
| Sodium Ethylate | 72 |
| Solution, 10, 13; Solutions | 71 |
| Solutions Percentage | 126 |
| Spearmint, Oil of | 77 |
| Spermaceti | 49 |
| Spirits of Nitre | 96 |
| Spirits | 13, 96 |
| Squill | 91 |
| Stavesacre | 96 |
| Starch | 35 |
| Strammonium | 96 |
| Storax | 97 |
| Strophanthin | 97 |

| | |
|---|---|
| Strychnine | 75 |
| Subscription | 116 |
| Suet | 93 |
| Sugar | 89 |
| Sulphonal | 97 |
| Sulphur | 97 |
| Sulphur Iodide | 97 |
| Sulphuric Acid | 29 |
| Sulphuric Ether | 30 |
| Sulphurous Acid | 29 |
| Sumbul | 98 |
| Superscription | 116 |
| Suppositories | 14, 98, 110, 131 |
| Sweet Spirit of Nitre | 96 |
| Synergists | 19 |
| Syrups | 14, 98 |
| Tabellæ | 16 |
| Tablets | 14, 16, 98 |
| Tamarinds | 98 |
| Tampons | 16 |
| Tannic Acid | 29 |
| Tannin | 29 |
| Tar | 83 |
| Tartar Emetic | 36 |
| Tartaric Acid | 29 |
| Terebene | 98 |
| Terpine Hydrate | 107 |
| Theobroma, Oil of | 78 |
| Theobromine | 107 |
| Therapeutics | 6 |
| Thymol | 99 |
| Thyroid | 99 |
| Tinctures | 14, 99 |
| Tolerance | 17 |
| Toughened Caustic | 37 |
| Tragacanth | 99 |
| Trinitrin | 72, 98 |
| Trituration | 11 |
| Trochiscus | 100 |
| Tuberculin | 107 |
| Turpentine, Oil of | 77 |
| Turpentine, Canada | 98 |
| Urotropin, see Hexamethylenamina | |
| Valerian | 100 |
| Vaseline, see Paraffinum Molle | 81 |
| Veratrine | 100 |
| Vinegars | 11 |
| Virginian Prune Bark | 86 |
| Vocabulary | 120 |
| Warming Plaster | 46 |
| Water, 36; Waters | 11, 36 |
| Weights and Measures | 6 |
| White Precipitate | 65 |
| Wines | 14, 101 |
| Wintergreen, Oil of; see Methyl Salicylicum | 105 |
| Witch-hazel | 63 |
| Wool Fat | 30 |
| Yellow Wash | 66 |
| Zinc | 101 |

CPSIA information can be obtained
at www.ICGtesting.com
Printed in the USA
BVHW090105211118
533638BV00011B/485/P

9 781331 950882